PYRIDINE

Edited by **Pratima Parashar Pandey**

Pyridine
http://dx.doi.org/10.5772/intechopen.71351
Edited by Pratima Parashar Pandey

Contributors

Amer Al Abdel Hamid, Yoshio Hamada, Satyanarayan Pal, Shigerus Satoh, Pratima Parashar

Notice

Statements and opinions expressed in the chapters are these of the individual contributors and not necessarily those of the editors or publisher. No responsibility is accepted for the accuracy of information contained in the published chapters. The publisher assumes no responsibility for any damage or injury to persons or property arising out of the use of any materials, instructions, methods or ideas contained in the book.

First published in London, United Kingdom, 2018 by IntechOpen
IntechOpen is the global imprint of INTECHOPEN LIMITED, registered in England and Wales, registration number: 11086078, The Shard, 25th floor, 32 London Bridge Street
London, SE19SG – United Kingdom
Printed in Croatia

British Library Cataloguing-in-Publication Data
A catalogue record for this book is available from the British Library

Additional hard copies can be obtained from orders@intechopen.com

Pyridine, Edited by Pratima Parashar Pandey
p. cm.
Print ISBN 978-1-78923-422-0
Online ISBN 978-1-78923-423-7

Meet the editor

Dr. Pratima Parashar Pandey, Phd, has been an academician and a scientist in the field of Materials Science and Nanotechnology for the last 30 years. She has published about 26 papers in cited journals. She has also written a chapter "Silver Particulate Films on Softened Polymer Composite" in the book *Applications of Calorimetry in a Wide Context - Differential Scanning Calorimetry, Isothermal Titration Calorimetry and Microcalorimetry*, Intech Publication. She has also written a chapter "Nanobiomaterials in Antimicrobial Therapy" in a book *Recent Biopolymers* published by Intech Publication. She has been a reviewer, a technical program member, and an invited speaker for many international and national conferences. This is her first assignment as an editor for the book titled *Pyridine*, Intech publication.

Contents

Preface

In the fast-growing world, for the ready and proper use of new revolutionized recent researches, developments and their applications in human life need to be complied. This book is an effort to do so, designed and styled in order to give researchers all-in-one-place about pyridine.

The book starts with wide structural analysis of pyridine, which is the base for its huge applications. Pyridines are used in industry and are also present in the form of pyridine moieties found in many natural products, such as vitamins, coenzymes, alkaloids, many drugs and pesticides.

The second chapter discusses the effects of substitution and substituent modification on electron density enrichment of the pyridine nitrogen and thus its effectiveness as an electron donor. The investigations provide insights into promoting and qualifying pyridine-based chelates to be good stabilizers for metal ion complexation in coordination chemistry.

The third chapter focuses on the chemistry as well as on various biological activities associated with pyridine ring system. Pyridine skeleton forms an essential part of various medicinal agents and represents a vast range of biological activities such as antimicrobial, antiviral, antihistaminic, anti-inflammatory, analgesic, fungicidal, herbicidal, insecticidal, antitumor, anticancer, CNS stimulant, analgesic, local anesthetic, and other activities. From the medicinal chemistry point of view, the nitrogen-containing heterocycles are more remarkable because these modify the electron distribution inside the scaffold leading to an alteration of the physical and chemical properties of the compounds.

The fourth chapter looks into the ligating capability of basic pyridine unit to transition metal ions. Pyridine owing to its Lewis basic character rooted in its nitrogen lone pair qualifies as ligand for transition metals and is able to form metal complexes across the metals in periodic table. It is usually a weak monodentate ligand having capability to bind metal in different proportions to produce range of metal complexes.

The fifth chapter explores the possibility to develop pyridine-*di*-carboxylic acids (PDCAs) and pyridine-*mono*-carboxylic acids (PCAs) as novel flower care agents as well as growth-promoting agents, which will be used for vegetable cultivation. Further, these acids stimulate root and shoot growth of lettuce, carrot, and rice seedlings.

I sincerely thank all the scientists whose works constitute this book.

I am thankful to IntechOpen for giving me the opportunity to edit this book. Despite best of my efforts, some shortcomings may have occurred, for which, I tender my sincere apology.

I wish all the best to esteemed readers.

Dr. Pratima Parashar Pandey
Professor, Nanotechnology
IILM College of Engineering & Technology
Greater Noida, India

Introduction

Chapter 1

Introductory Chapter: Pyridine

Pratima Parashar Pandey

Additional information is available at the end of the chapter

http://dx.doi.org/10.5772/intechopen.77969

1. Introduction

Pyridine (C_5H_5N), an aromatic compound where all the pi electrons are shared by a ring, forms one continuous circle of electrons besides the alternate double bonds shared by every atom on the circle. Pyridine is a unique type with nitrogen on the ring to provide a tertiary amine by undergoing reactions such as alkylation and oxidation. Amine is responsible for the slight dipole on the ring because electrons are drawn more toward the nitrogen being electronegative (lone pair electrons on the nitrogen) than other atoms in the ring. The H nuclear magnetic radiation (H-NMR) shows three signals at the ortho, meta, and para positions on the molecule in respect of three different chemical shifts. These chemical shifts are the result of the different electron densities for each of these atoms. As a result, this is not stable as other aromatic compounds (**Figure 1**).

Pyridine, a liquid similar to water, can easily mix with water and other organic solvents. This property is useful for making various products such as medicines, vitamins, food flavorings, pesticides, paints, dyes, rubber products, adhesives, waterproofing fabrics, and nitrogen-containing plant products. This nature of pyridine further makes it to be used as a precursor for many agrochemicals and pharmaceuticals. Hence, pyridine and its derivatives have significant applications in various fields, especially in the medicinal area.

All these properties of pyridine make it worthwhile to have a full overview about pyridine and its derivatives with recent researches in one place for potential researchers. This book provides a range of chapters which touches each and essential aspect of pyridine. It is the electronegative nature of pyridine in the backdrop responsible for the formation of all its derivatives. The synthesis of derivatives using pyridine has the biological activities and vast applications. Pyridine ring is also advantageous in the formation of discontinuous silver films on polymer composite by using its one of the derivative poly (4-vinyl pyridine) (P4VP).

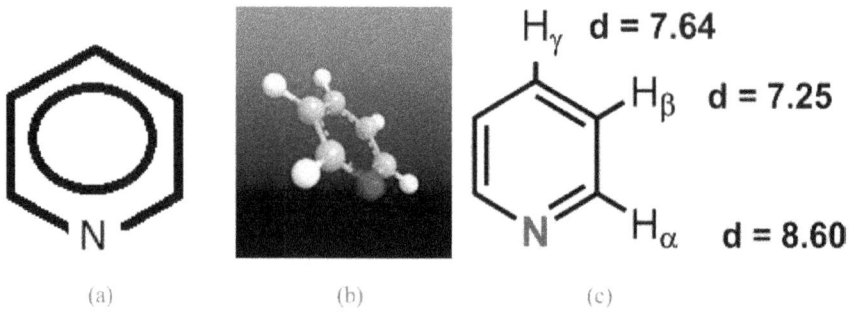

Figure 1. (a) The double bonds in pyridine are shared between all of the atoms in the circle so that they can be drawn as a circle instead of individual bonds. (b) Pyridine. (c) The hydrogen atoms on pyridine have three distinct chemical shifts. Pyridine shares the electrons in the entire circle, but the nitrogen draws more electrons to it, causing a slight dipole to form.

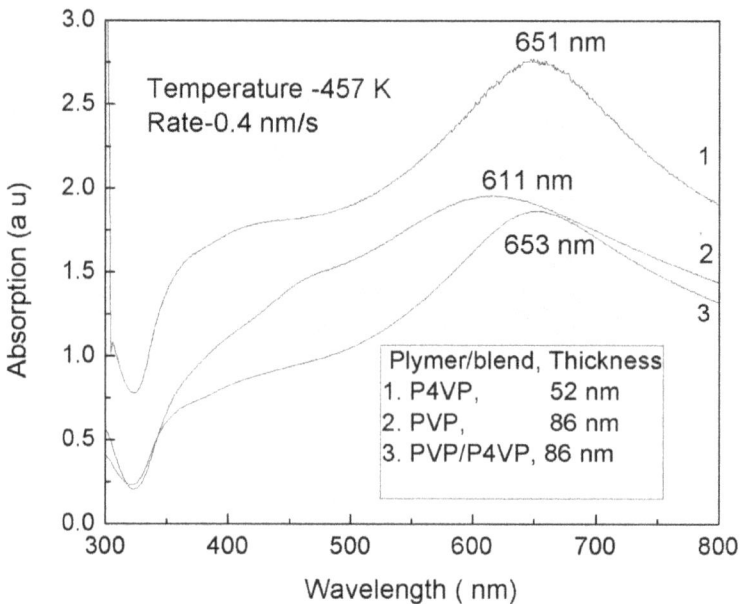

Figure 2. Optical absorption spectra of P4VP, PVP, and their composite.

2. Pyridine a link to Silver nanocomposite

Nanosized silver particles are well known to be used in many bactericidal and antimicrobial applications [1–4]. Antimicrobial property is most important to prevent a wound from microorganism for a long healing process. A lot of disinfectant agents have been introduced to wound dressings. Among them, using Ag NPs is the most advanced technology as its high efficiency has been proven. Preparing silver nanoparticles by toxic precursor chemicals produces hazardous by-products and contaminated Ag NPs, and hence, these must be produced with safe and clean methods.

Polyvinyl pyrrolidone (PVP) attracts considerable attention because of its excellent chemical and physical properties making it an excellent material as a coating or as an additive to different materials. It needs to be uniformly incorporated with Ag NPs. If PVP alone is used as host polymer agglomerated, nonuniform silver films are formed. The pyridine in P4VP because of the lone pair of the electron in nitrogen helps in the formation of nonagglomerated, uniform films on PVP/P4VP (50:50) [5] composites. The optical absorption of spectra shows the embedment of silver nanoparticles in the composite (**Figure 2**).

Author details

Pratima Parashar Pandey

Address all correspondence to: pratimaparashara@rediffmail.com

IILM College of Engineering and Technology, Greater Noida, India

References

[1] Irwin P, Martin J, Nguyen L-H, He Y, Gehring A, Chen Chin-Yi. Antimicrobial activity of spherical silver nanoparticles prepared using a biocompatible macromolecular capping agent: Evidence for induction of a greatly prolonged bacterial lag phase. Journal of Nanobiotechnology. 2010;**8**:34. DOI: 10.1186/1477-3155-8-34

[2] Bryaskova R, Pencheva D, Nikolov S, Kantardjiev T. Synthesis and comparative study on the antimicrobial activity of hybrid materials based on silver nanoparticles (AgNPs) stabilized by polyvinylpyrrolidone (PVP). Journal of Chemical Biology. 2011;**4**:185-191. DOI: 10.1007/s12154-011-0063-9

[3] Basri H, Ismail A F, Aziz M. Assessing the effect of PVP of various molecular weight (MW) in PES-Ag membranes: Antimicrobial study using E.Coli. Journal of Science and Technology. 2011;**3**(2):59-67

[4] Xiong J, Ghori M Z, Henkel B, Strunskus T, Schürmann U, Deng M, Kienle L, Faupel F. Tuning silver ion release properties in reactively sputtered Ag/TiOx nanocomposites. Applied Physics A. 2017;**123**(7):470-481. DOI: 10.1007/s00339-017-1088-x

[5] Parashar P: Synthesis of Silver Nanocomposite with Poly (vinylpyrollidone) and Poly(4-vinylpyridine) for Antimicrobial Activity. Advanced Materials Research. 2013;**772**:9-14. DOI: 10.4028/www.scientific.net/AMR.772.9

Pyridine: In The Medicinal World

Role of Pyridines in Medicinal Chemistry and Design of BACE1 Inhibitors Possessing a Pyridine Scaffold

Yoshio Hamada

Additional information is available at the end of the chapter

http://dx.doi.org/10.5772/intechopen.74719

Abstract

Pyridine is a unique aromatic ring. Although pyridines are used industrially, pyridine moieties are present in many natural products, such as vitamins, coenzymes, and alkaloids, and also in many drugs and pesticides. Pyridine moieties are often used in drugs because of their characteristics such as basicity, water solubility, stability, and hydrogen bond-forming ability, and their small molecular size. Because pyridine rings are able to act as the bioisosteres of amines, amides, heterocyclic rings containing nitrogen atoms, and benzene rings, their replacement by pyridine moieties is important in drug discovery. Recently, we synthesized a series of BACE1 inhibitors by in silico conformational structure-based drug design and found an important role of pyridine moiety as a scaffold. In this chapter, we describe the important role of pyridines in medicinal chemistry and the development of β-secretase inhibitors possessing a pyridine scaffold for the treatment of Alzheimer's disease.

Keywords: Alzheimer's disease, BACE1 inhibitor, bioisostere, drug design, *in silico* conformational structure-based design

1. Introduction

Alzheimer's disease (AD) is the most common cause of dementia. AD is characterized by progressive intellectual deterioration. In 1906, Alois Alzheimer, a psychiatrist and a neuropathologist, reported on a 51-year-old female at the Frankfurt Asylum. The patient showed strange behavioral symptoms and loss of short-term memory, which was later called "AD." The cause of AD has only been clarified relatively recently, and there have been no therapeutic agents since that first report by Dr. Alzheimer over 100 years ago. Recently, the development of

IntechOpen

many drug candidates based on the amyloid hypothesis has been reported. β-Secretase (BACE1; β-site amyloid precursor protein-cleaving enzyme 1) is a promising molecular target for the development of anti-Alzheimer's drugs. BACE1 triggers the formation of the amyloid β (Aβ) peptide that is the main component of the senile plaques found in the brain of AD patients. We designed a series of peptidomimetic inhibitors possessing a substrate transition-state analog. We followed this with the design of nonpeptidic BACE1 inhibitors possessing a pyridine scaffold, using an approach based on a conformer of the docked ligand in the target biomolecule—the *"in-silico* conformational structure-based design." In this process, we noticed an important and third role of pyridines in medicinal chemistry. Pyridines are contained in many natural products, such as vitamins, coenzymes, and alkaloids. Pyridine moieties are often used in drugs and pesticides because of characteristics that include basicity, water solubility, stability, hydrogen bond-forming ability, and small molecular size. In this chapter, the conventional roles of pyridine in medicinal chemistry are described. We also introduce another role using our example regarding the design of BACE1 inhibitors.

2. Conventional roles of pyridine in medicinal chemistry

Pyridine rings are present in many natural products including vitamins such as niacins and vitamin B_6, coenzymes such as nicotinamide adenine dinucleotide (NAD), and alkaloids such as trigonelline. Trigonelline is an alkaloid that is the product of niacin metabolism. Many drugs and pesticides contain a pyridine moiety. Examples include antimicrobial agents, antiviral agents, antioxidants, antidiabetic agents, anti-malarial agents, anti-inflammatory agents, psychopharmacological antagonists, and antiamoebic agents [1]. These pyridine moieties play critical roles in medicinal chemistry because of their abovementioned characteristics. One role of pyridine in medicinal chemistry is to improve water solubility because of its weak basicity. Although many drugs and pesticides possessing a pyridine ring had been designed for improved water solubility, this improvement is often pH-dependent. For example, the sulfa drug, sulfapyridine 1 (**Figure 1A**) has good antibacterial activity and water solubility under acidic conditions, but there is a risk of crystallization in the bladder or urethra, which leads to pain or blockage of the urethra. The conjugate of sulfapyridine and 5-aminosalicylic acid by an azo bond is the compound called sulfasalazine 2 (**Figure 1A**). It displays good water solubility and is used in clinical practice in the treatment of rheumatoid arthritis, ulcerative colitis, and Crohn's disease [2]. When the parent compound cannot be substituted with a pyridine ring, there is an alternative solution—water-soluble prodrug. Water-soluble prodrug 3b (isavuconazonium sulfate) of an antifungal agents is shown in **Figure 1B** [3]. Prodrug 3b hydrolyzed by an esterase to release an intermediate 4b. Prodrug 4b can spontaneously release the parent drug, isavuconazole 5, in physiological conditions. Prodrug 3b, which possesses a pyridine ring, displays good water solubility (>100 mg/mL) compared to prodrug 3a (>10 mg/mL) that lacks pyridine ring. Prodrug 3b was approved as an oral medicine by the United States Food and Drug Administration (FDA) in 2015.

A

1 (sulfapyridine)

2 (salazosulfapyridine)

B

3a (X = CH)
3b (X = N) RO0098557 (Isavuconazonium sulfate)

esterase

intermediate   **4a** (X = CH)
4b (X = N)

spontaneous

5 RO0094815 (Isavuconazole)

+ CH₃CHO

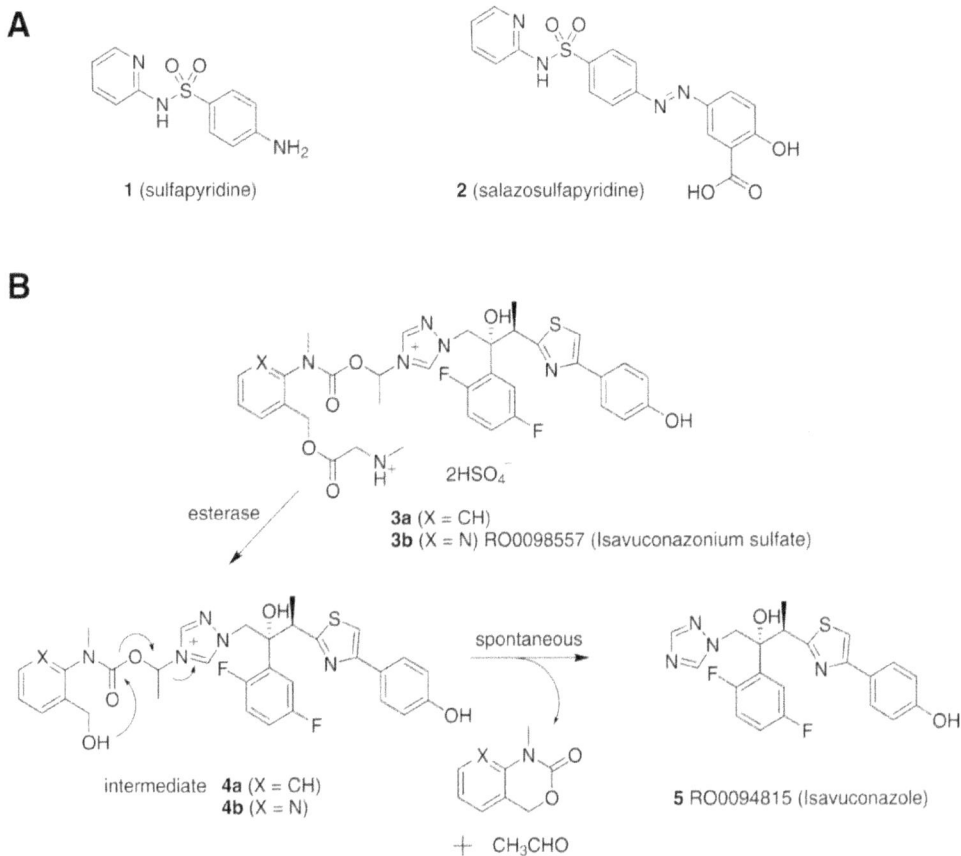

Figure 1. (A) Drugs with improved water solubility. (B) Isavuconazonium sulfate, a water-soluble prodrug possessing a pyridine ring.

Bioisosteres have an important role in the pyridine ring for medicinal chemistry [4]. Bioisosteres are functional or atomic groups with similar physiochemical properties to the parent functional/atomic groups.

Compounds associated with them exhibit similar biological or physiochemical properties as the parent compound. In medicinal chemistry, a portion of a candidate drug is replaced with other functional/atomic groups with the goal of improving drug efficacy, *in vivo* stability, oral absorption, membrane permeability, and absorption, distribution, metabolism and excretion (ADME). Among the approaches used for drug discovery research, the modification of drug candidates by their corresponding bioisosteres is the first choice in drug design studies. Because pyridine is a unique aromatic ring that features a small molecular size, weak basicity, and good stability, pyridine rings had been used as the bioisostere for other heterocyclic aromatic rings, benzene rings, amides, and amines [4]. Especially, pyridines are often replaced with monocyclic aromatic rings, such as benzenes, imidazoles, pyrrole, and oxazole rings, because of their same molecular size as the pyridine ring. Some drugs with a pyridine ring as a bioisostere of imidazole and benzene ring are presented in **Figure 2**. Histamine has an amino group and an imidazole ring. Thus, histamine receptor antagonists with the respective bioisosteres of the amino group and an imidazole ring have been designed. Cimetidine **6** (**Figure 2A**) is an H_2 receptor antagonist, which has an imidazole ring and a guanidine derivative as the analog of an amino group [5]. The H_2 receptor belongs to the rhodopsin-like family of G protein-coupled receptors. Because the H_2 receptor stimulates gastric acid secretion, its antagonists such as cimetidine are used in the treatment of heartburn and peptic ulcers. Mepyramine **7** (**Figure 2A**) has a pyridine ring as a bioisostere of the imidazole ring. The compound has histamine H_1 receptor antagonist activity [5]. Histamine receptor H_1 is expressed in smooth muscles, on vascular endothelial cells, in the heart, and in the central nervous system. H_1 receptor antagonists are used as antiallergy drugs. Some histamine antagonists in which imidazole ring of the ligand is replaced with a bulky aromatic ring such as doxepin [6] display H_1 receptor antagonist activity, indicating that this ligand site appears to decide the affinity toward the histamine receptor subtypes.

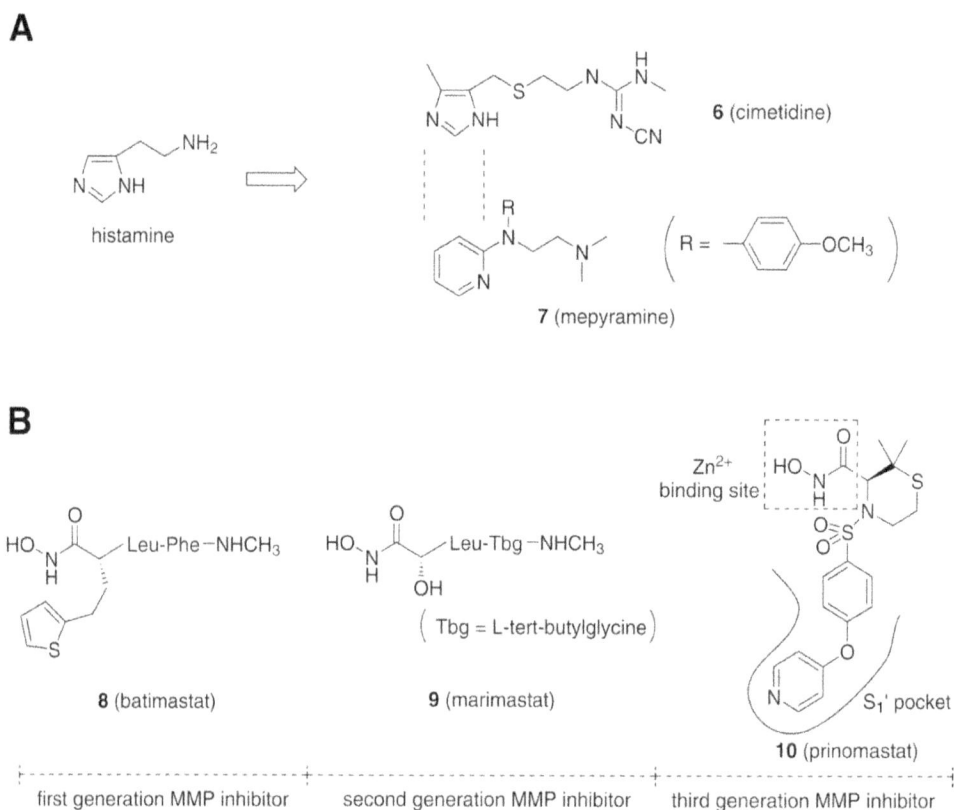

Figure 2. (A) Histamine and histamine receptor antagonists and (B) matrix metalloproteinase (MMP) inhibitors.

Matrix metalloproteinases (MMPs) are calcium- and zinc-containing endopeptidases that have diverse roles in cell behaviors including cell proliferation, migration, and differentiation. Some MMP family subtypes, which include MMP2, MMP3, and MMP9, can degrade the extracellular matrix, resulting in the accelerated infiltration and migration of cancer cells. Inhibitors of some MMP subtypes had been reported for anticancer activity. Most MMP inhibitors have a hydroxamic acid that can bind to the zinc-containing sites of MMPs [7]. The second-generation MMP inhibitors such as batimastat **8** have a more potent inhibitory activity than that of the first-generation MMP inhibitors such as marimastat **9**. (**Figure 2B**), but their selectivity against MMP subtypes is insufficient. The third-generation MMP inhibitor prinomastat **10** has a pyridine ring at the P_1' position, which improves the selectivity to MMP2 and MMP9. A deep hydrophobic pocket corresponding to the S_1' sites is located near the zinc-binding site of MMP, and an alkyl or phenyl groups of amino acids of inhibitors can bind to the S_1' pocket. Researchers at Shionogi & Company Limited reported that the biphenyl group of a third-generation MMP inhibitor, BPHA, can interact tightly with the deep S_1' pocket [8]. The pyridine ring of prinomastat appears to behave as a bioisostere of the benzene ring. As stated earlier, pyridines had been used in medicinal chemistry because of their unique properties, such as weak basicity, water solubility, *in vivo*/chemical stability, hydrogen bond-forming ability, or small molecular size.

3. Design of BACE1 inhibitors

3.1. Pathology of AD and design of peptidomimetic inhibitors

AD is the most common cause of dementia. Its cause has been unclear. A breakthrough was made through the genetic study of some familial AD (FAD) patients with a mutation of the gene encoding amyloid precursor protein (APP) or presenilin gene. As these mutations caused an increase in Aβs that are the main components of senile plaques in the brain of patients with AD, it indicates their involvement in the pathogenesis of AD [9–12]. Aβs are produced from APP by two processing enzymes, β-secretase (BACE1; β-site APP-cleaving enzyme 1) and γ-secretase, which are potential molecular targets for anti-AD drugs [13–16]. The cleavage sites of APP are shown in **Figure 3A**. The full-length APP (APP770) and its isoforms, APP695 and APP751, result from the alternative splicing of its mRNA. BACE1 is a type I transmembrane aspartic protease with 501 amino acids, which triggers Aβ formation in the rate-limiting first step by cleaving at the N-terminus (β-site) of the Aβ domain of APP. Next, the aspartic protease, γ-secretase, cleaves at the C-terminus of the Aβ domain, releasing Aβs that consists mainly of two molecular species, $A\beta_{1-42}$ and $A\beta_{1-40}$. γ-Secretase cleaves two cleavage sites "γ-sites" forming $A\beta_{1-40}$ and $A\beta_{1-42}$. Two processing enzymes, BACE1 and γ-secretase, are categorized as aspartic proteases. They have an acidic optimum pH. Furthermore, BACE1 and γ-secretase, and their substrate, APP, are located in the same intracellular granules, such as endosomes and the trans-Golgi network, which have an acidic environment, suggesting that Aβs are produced in these locations [17]. $A\beta_{1-42}$ displays more potent neurotoxicity and aggregation behavior than $A\beta_{1-40}$ and appears to be critical in the pathogenesis of AD. By contrast,

A

B

11 (OM99-2, BACE1 *K*i = 1.6 nM)   **12** (OM00-3, BACE1 *K*i = 0.3 nM)

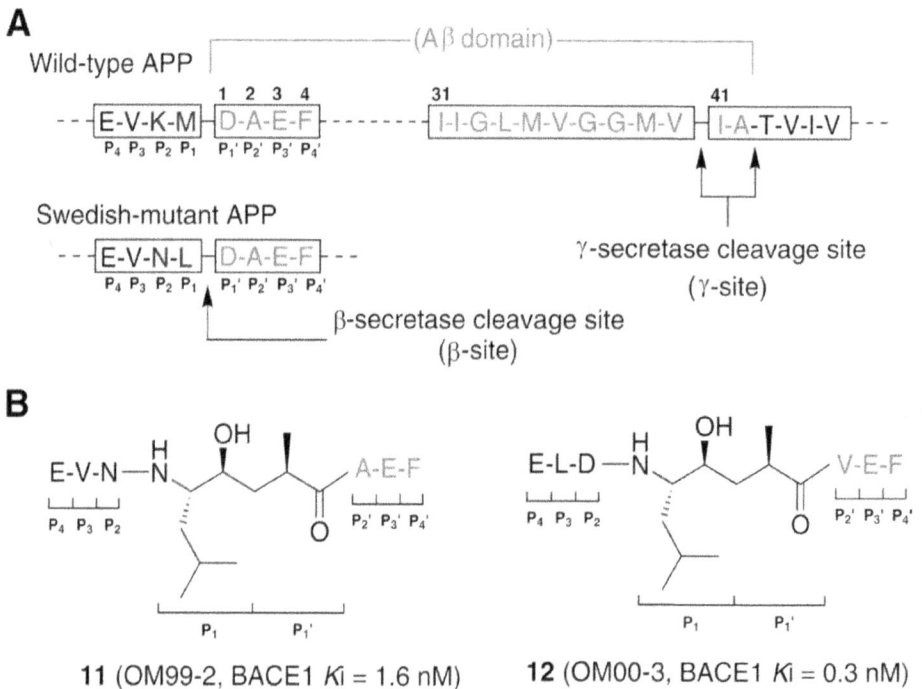

Figure 3. (A) Amyloid precursor protein (APP) and its cleavage site and (B) peptidomimetic BACE1 inhibitors designed on the basis of Swedish-mutant APP sequence by Ghosh et al.

α-secretase is a disintegrin and metalloprotease (ADAM) family metalloprotease, for example, ADAM9, ADAM10, and TNF-α-converting enzyme (TACE, also known as ADAM17), which cleaves APP at the α-site between Lys16 and Leu17 in the Aβ domain [17]. A homolog enzyme of BACE1, BACE2, cleaves at two sites (θ-sites) between Phe19 and Phe20, and between Phe20 and Ala21 in the Aβ domain [18]. Because the α-site and θ-sites are located at the center of the Aβ domain, their cleavage does not lead to Aβ production. According to the amyloid hypothesis, BACE1 and γ-secretase are the molecular targets for anti-AD drugs. However, because γ-secretase can cleave other single-pass transmembrane proteins *in vivo* such as Notch, which plays a critical role in cell differentiation, its inhibition appears to lead to serious side effects. The fact that BACE1 knockout transgenic mice can survive normally has provided a promising road map, in which BACE1 is a molecular target for the development of AD drugs [19].

As BACE1 is an aspartic protease, early BACE1 inhibitors are peptidomimetic with a substrate transition-state analog. They were designed on the basis of an inhibitor design approach as well as other aspartic proteases such as renin and human immunodeficiency virus protease [20–26]. Many mutations in the APP gene that affect Aβ formation, $A\beta_{1-42}/A\beta_{1-40}$ ratio or Aβ toxicity have been reported. Among them, the Swedish mutation (K670 N, M671 L double mutation, **Figure 3A**) around the β-site induces β-cleavage by BACE1, increasing the $A\beta_{1-42}$ and $A\beta_{1-40}$ levels in the brains of AD patients. Because the Swedish-mutant APP is cleaved faster than the wild-type APP, early BACE1 inhibitors were designed on the basis of the

Swedish-mutant APP amino acid sequence. In 1999, Sinha et al. at Elan Pharmaceuticals succeeded in purifying BACE1 from the human brain using a transition-state analog based on the Swedish-mutant sequence and cloned the BACE1 enzyme [16]. In 2000 and 2001, Ghosh and Tang described the potent inhibitors, compounds **11** (OM99–2, Ki = 1.6 nM) and **12** (OM00–3, Ki = 0.3 nM) with a transition-state analog corresponding to a dipeptide unit at the P_1–P_1' positions of APP (**Figure 3B**) and provided the first X-ray crystal structure (PDB ID: 1FKN) of a complex between recombinant BACE1 and the inhibitor OM99–2 [27–30].

We have also reported a series of peptidomimetic BACE1 inhibitors possessing a norstatine-type transition-state analog, phenyl norstatine (Pns: (2R,3S)-3-amino-2-hydroxy-4-phenylbutyric acid), at the P_1 position as shown in **Figure 4** [31–38]. These inhibitors have a Glu bioisostere at the P_4 position and a C-terminus anilide substituted by an acidic group corresponding to the Asp residue at the P_1' position of the APP sequence. Among the compounds, **13** (KMI-429, IC_{50} = 3.9 nM) effectively inhibits BACE1 activity in cultured cells and significantly reduces Aβ production *in vivo* when directly administered into the hippocampi of APP transgenic and wild-type mice [31, 32]. The most potent inhibitor, compound **14** (KMI-684, IC_{50} = 1.2 nM) features two carboxylic acid residues of KMI-429 at the P_1' position that have been replaced with their bioisostere, a tetrazolyl ring [33]. Compound **15** (KMI-574, IC_{50} = 5.6 nM), which possesses a 5-fluoroortyl group in the N-terminus residue, displays improved inhibition in cultured cells because of improved cell membrane permeability [34].

3.2. Design of nonpeptidic BACE1 inhibitors with a pyridine scaffold

We designed and synthesized nonpeptidic BACE1 inhibitors from our peptidic BACE1 inhibitors **13–15** as lead compounds [20–25]. Researchers at MSD, Elan, and Pfizer, and Gosh et al. reported a series of BACE1 inhibitors possessing an isophthalic scaffold at the P_2 position [23, 25].

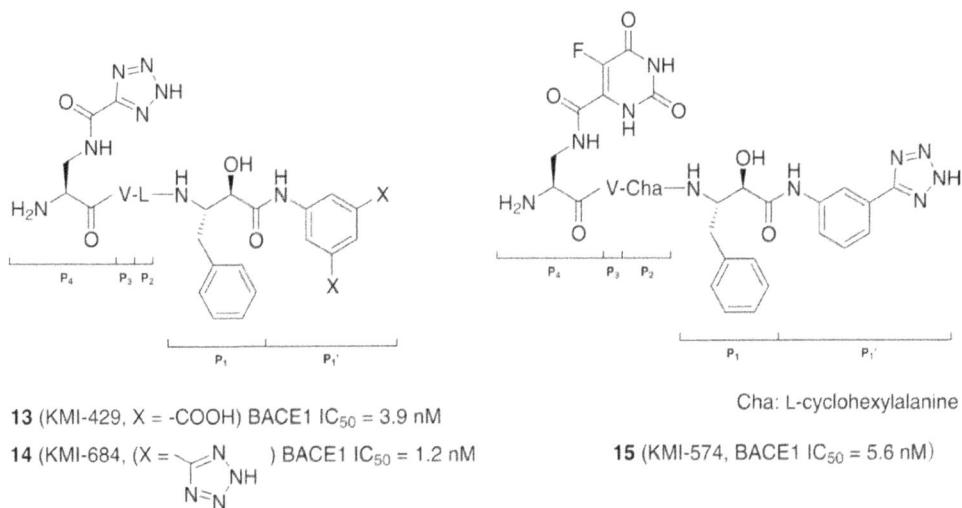

13 (KMI-429, X = -COOH) BACE1 IC_{50} = 3.9 nM

14 (KMI-684, (X = ...)) BACE1 IC_{50} = 1.2 nM

Cha: L-cyclohexylalanine

15 (KMI-574, BACE1 IC_{50} = 5.6 nM)

Figure 4. Peptidomimetic BACE1 inhibitor with a norstatine-type substrate transition-state analog.

The inhibitors formulated by Elan and MSD researchers, compounds **16** and **17**, respectively, are shown in **Figure 5**. Because the distance between the flap domain and the cleft domain that form the S_2 pocket of BACE1 is narrow, a planar aromatic ring such as an isophthalic scaffold can closely dock in the S_2 pocket of BACE1. Hence, we designed BACE1 inhibitor **18** with an isophthalic scaffold at the P_2 position [39]. However, compound **18** showed a low inhibitory activity (BACE1 inhibition 55% at 2 μM). Our next innovation involved the S_3 sub-pocket located behind the active site of BACE1. A docking simulation study between

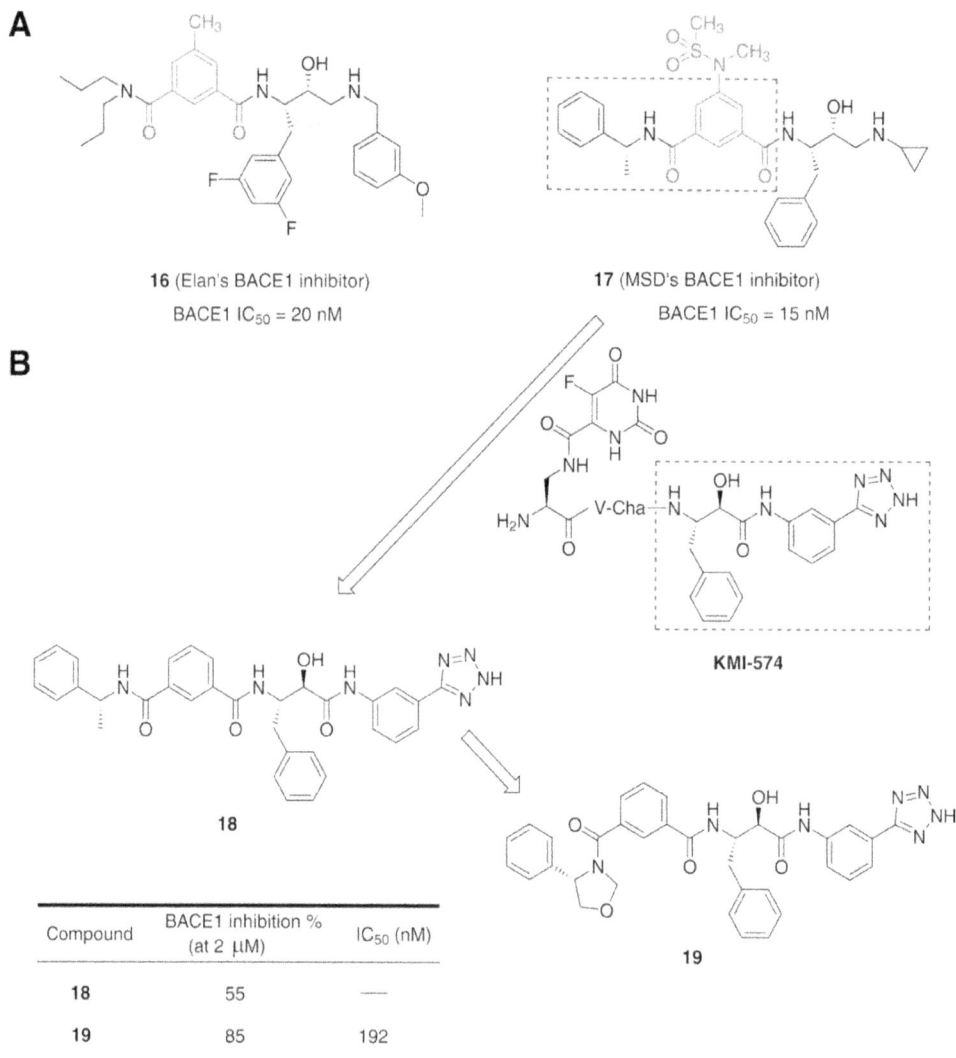

A

16 (Elan's BACE1 inhibitor)
BACE1 IC$_{50}$ = 20 nM

17 (MSD's BACE1 inhibitor)
BACE1 IC$_{50}$ = 15 nM

B

KMI-574

18

19

Compound	BACE1 inhibition % (at 2 μM)	IC$_{50}$ (nM)
18	55	—
19	85	192

Figure 5. (A) BACE1 inhibitors with an isophthalic scaffold reported by Elan Pharmaceuticals/Pharmacia and MSD and (B) design of isophthalic-type BACE1 inhibitors using a norstatine-type transition-state analog of KMI-574.

inhibitor **18** and BACE1 revealed that the P_3-phenyl group of inhibitor **18**, which interacts with the S_3 sub-pocket, adopts a folding structure against the P_2-isophthalic scaffold. We envisioned and designed an inhibitor that possessed a folding structure and synthesized inhibitor **19**. This compound featured a five-membered ring, oxazolidine, at the P_3 position in order to fix the folding pose between the P_2-phenyl group and P_3-isophthalic scaffold. Our premise was that the oxazolidine ring fixes the direction of the phenyl ring at the P_3 position, so the P_3-phenyl ring might be able to bind closely to the S_3 sub-pocket of BACE1. Inhibitor **19** showed moderate inhibitory activity (BACE1 inhibition 85% at 2 μM, IC_{50} = 192 nM).

Next, we focused on a proton of the P_2-isophthalic ring of inhibitor **19**. We demonstrated van der Waals repulsion between the proton on the isophthalic ring at the P_2 position and the five-membered ring at the P_3 position in inhibitor **19** docked at the active site of BACE1. We focused on the steric-hindered interaction between the P_3-phenyl group and a proton on the P_2-isophthalic ring of a virtual inhibitor (**Figure 6**), which seemed to restrict its configuration. We calculated the steric energies in the respective conformers around the bond of the P_3 amide and P_2-isophthalic ring of the virtual inhibitors as shown in **Figure 6**. Using an approach based on a conformer of the docked inhibitor in BACE1 (the *in silico* conformational structure-based design) [39, 41], we adopted a pyridinedicarboxylic scaffold as a P_2 moiety, which lacked a proton from the isophthalic ring. Whereas the conformer of the P_2-isophthalic virtual inhibitor with the same dihedral angle to the conformer docked in BACE1 showed a high steric energy, the stable conformer of the virtual inhibitor with a P_2-pyridinedicarboxylic scaffold showed the same dihedral angle to that docked in BACE1 because of the lack of a proton on the pyri-

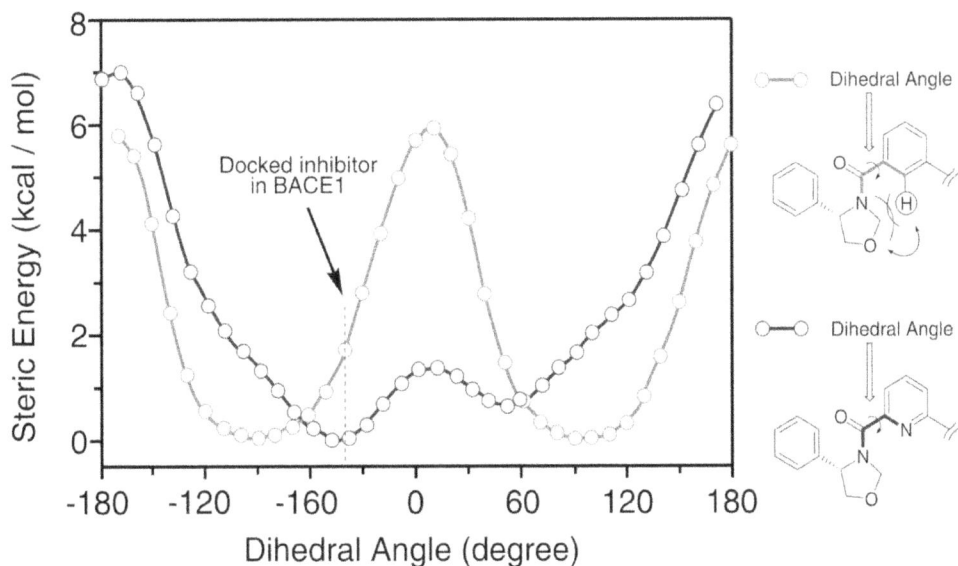

Figure 6. In-silico conformational structure-based design of BACE1 inhibitor possessing a pyridine scaffold at the P_2 position.

dine's amine. Inhibitor **20** with a pyridinedicarboxylic scaffold was designed and synthesized and showed improved inhibitory activity (BACE1 inhibition 93% at 2 μM, IC_{50} = 140 nM) compared to **19** (**Table 1**). A docking simulation demonstrated that inhibitor **20** could adopt a stable folding structure having the same dihedral angle between the P_3-amide and P_2-isophthalic ring to the conformer docked in BACE1. Thus, we could design a potent BACE1 inhibitor (compound **20**) using the computational approach based on the conformer docked in BACE1. However, **20** still showed a lower inhibitory activity than our peptidic inhibitors compared to lead compounds **13–15** (IC_{50} = 1.2–5.6 nM). There is room for further optimization of these inhibitors.

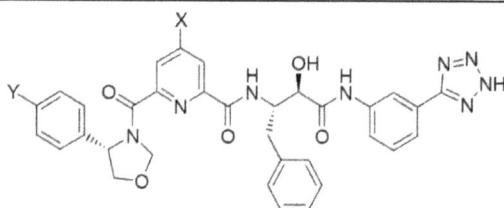

Compound	X	Y	BACE1 inhibition %		IC_{50} (nM)
			at 2 μM	at 0.2 μM	
20	-H	-H	93	63	140
21	-SO$_2$CH$_3$	-H	96	73	96
22	-OCH$_3$	-H	91	53	151
23	-OEt	-H	79	—	—
24	-OPr	-H	64	—	—
25	-SCH$_3$	-H	95	66	89
26	-SCH(CH$_3$)$_2$	-H	72	—	—
27	-N$_3$	-H	95	68	79
28	-NH$_2$	-H	78	53	—
29	-N(CH$_3$) SO$_2$CH$_3$	-H	75	—	—
30	-CH$_3$	-H	95	68	—
31	-Cl	-H	98	87	22
32	-Br	-H	99	88	15
33	-I	-H	99	86	24
34	-Cl	-F	99	91	13
35	-Br	-F	99	93	9
36	-I	-F	99	92	10

Table 1. BACE1 inhibitors with a pyridine scaffold.

3.3. Design based on quantum chemical interaction and electron donor bioisostere

The first reported coordinate set of crystal structure of BACE1-inhibitor (OM99–2) complex is 1FKN by Gosh et al. [27–29]. The P_2 moiety of the inhibitor interacts with the Arg235 side chain of BACE1 by hydrogen bonding in the crystal structure. We compared the publicly available X-ray crystal structures of BACE1-inhibitor complexes and discovered that most inhibitors did not interact with Arg235 by hydrogen bonding [41]. Surprisingly, the guanidino group of BACE1-Arg235 in most crystal structures, except 1FKN, showed the similar "flopping over" feature of the P_2 region of the inhibitors, and the nearest distances between the guanidino plane of Arg235 side chain and the P_2 region of the inhibitor showed similar values of approximately 3 Å. The P_2 moieties in many crystal structures that interact with the BACE1-Arg235 side chain are a methyl group, carbonyl oxygen atom, or aromatic ring. They appear to interact with the guanidine plane of Arg235 side chain by CH-π, O-π, or π-π stacking interactions. This suggests that the π-orbital on the guanidino plane can interact with the P_2 region of the inhibitors by a weak quantum force. The only exception was the interaction in the first reported X-ray crystal structure, 1FKN. Although the P_2 moiety of OM99–2 in the crystal structure of 1FKN appeared to interact with the BACE1-Arg235 side chain via hydrogen bonding, the P_2-moiety of OM00–3 that was structurally similar to OM99–2 interacted with the π-orbital on the guanidine plane of the BACE1-Arg235 side chain via O-π interaction (PDB ID: 1M4H). Many early BACE1 inhibitors that possess a hydrogen bond receptor at the P_2 position were designed using the 1FKN crystal structure. However, the hydrogen-bonding interaction between most of the inhibitors and the BACE1-Arg235 side chain was not shown in their crystal structures. For instance, inhibitor **17** that was synthesized by the MSD researchers interacted with the BACE1-Arg235 side chain via a CH-π interaction (PDB ID: 2B8L) as shown in **Figure 7A** (PDB ID: 2B8L). It is likely that the researchers designed an inhibitor that possessed an N-methyl-N-methanesulfonyl group at the P_2 position in anticipation of the hydrogen-bonding interaction between the sulfonyl oxygen atom and the BACE1-Arg235 side chain. However, the N-methyl group of the inhibitor interacted with the π-orbital on the guanidine plane of the BACE1-Arg235 side chain at a distance of 2.8 Å. As described earlier, most of the BACE1 inhibitors, except OM99–2, in the crystal structure 1FKN, interacted with the BACE1-Arg235 side chain by a weak quantum force such as stacking or σ-π interaction. The Arg235 side chain of the BACE1-OM99–2 complex (1FKN) assumed an exceptionally different pose to the other crystal structures. As many researchers have designed BACE1 inhibitors with a hydrogen bond receptor on the basis of the first reported crystal structure 1FKN, docking models using 1FKN will require further review. Furthermore, we found that the side chain of BACE1-Arg235 could move in concert with the inhibitor's size. The guanidino planes of BACE1-Arg235 in the crystal structures of most BACE1 complexes showed similar distances from the P_2 regions of the inhibitors regardless of their molecular size. This potentially posed a serious issue for a docking simulation for the drug discovery of BACE1 inhibitors. However, the BACE1-Arg235 side chain seems to have a restricted range of motion: the BACE1-Arg235 side chain slides sideways, not up and down, along the wall of the β-sheet structure that consists of four peptide strands behind the flap domain of BACE1. Therefore, the location of the BACE1-Arg235 side chain could be predicted by the inhibitor's size. We hypothesized that the role of the BACE1-Arg235 side chain is important for the

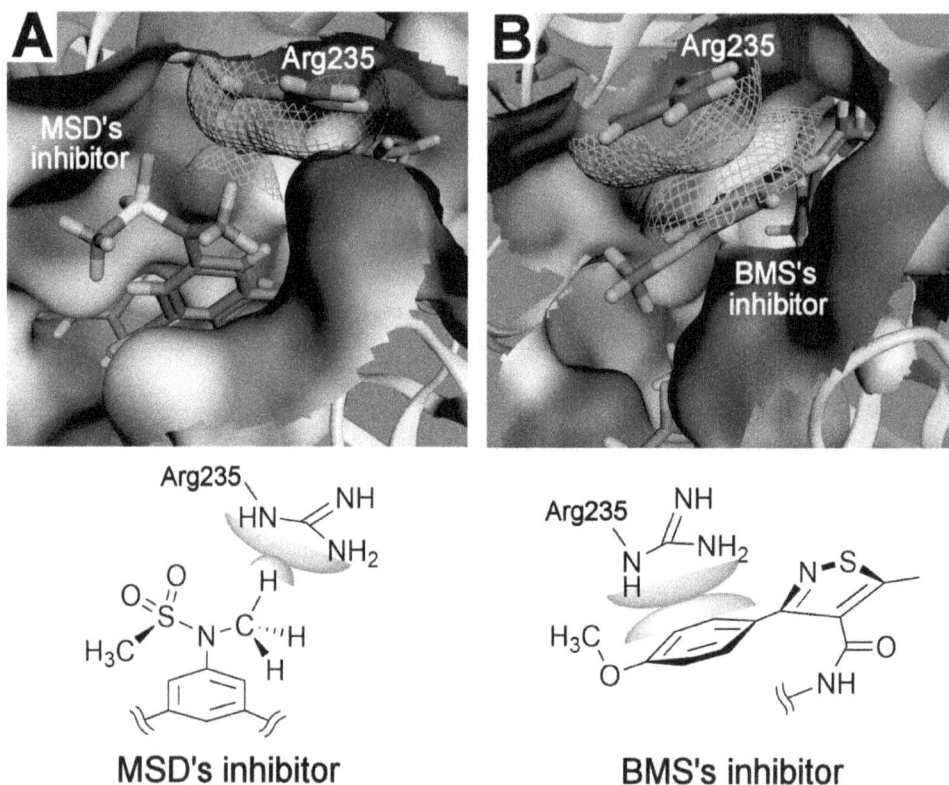

Figure 7. Interaction of BACE1 inhibitors with the Arg235 side chain of BACE1. (A) MSD's inhibitor 17 (PDB ID: 2B8L) and (B) BMS's inhibitor (PDB ID: 4FSL).

inhibitory mechanism of BACE1. The guanidine plane of Arg235 that can move in concert with the inhibitor's size appears to push down on the P_2 region of the inhibitor, which causes them to be affixed to the active site of BACE1 because of this "flop-over" mechanism by the BACE1-Arg235 side chain. Although a quantum chemical force, such as σ-π interaction, has a weaker binding energy than a hydrogen-bonding interaction, this "flop-over" mechanism permits a strong binding mode with the active site of BACE1.

In silico drug discovery using a docking simulation between a target biomolecule and drugs has provided important information. However, most docking simulation software involves mechanism/molecular dynamics (MM/MD) calculations based on classical Newtonian mechanics. Docking simulations using these calculations do not appear to estimate a weak quantum chemical interaction, such as stacking or σ-π interaction. The quantum chemical interactions that also involve other aromatic amino acids including phenylalanine (Phe), tyrosine (Tyr), and tryptophan (Trp) side chains seem to be approximately optimized using several descriptors based on classical mechanics in the docking simulation software that is based on MM/MD calculations. However, the software programs recognize arginine (Arg) as one of the charged amino acids, and the quantum chemical interactions involving an Arg side chain are unlikely to yield a reasonable output. Quantum chemical interactions involving a π-orbital of a guanidino group are common in proteins and play an important role in molecular recognition by proteins. Crowley et al. surveyed cation-π interactions in protein interfaces using the Protein Data Bank and the Protein Quaternary Structure server [42]. They evaluated the cation-π interactions using a variant of the optimized potentials for liquid simulations (OPLS) force field and found that approximately half of the protein-protein complexes and one-third of the homodimers contained at least one intermolecular cation-π pair. This finding indicates the significance of these interactions in molecular recognition because the occurrence rate of cation-π pairs in protein-protein interfaces is higher than that in homodimer interfaces, which are similar to the protein interior. Among them, the interactions between an Arg and a Tyr were found to be the most abundant. Moreover, 53% of them involved planar π-π stacking by the quantum chemical interaction between the guanidine group of an Arg residue and the aromatic ring of a Tyr residue. Researchers at Bristol-Myers Squibb Research (BMS) reported a series of BACE1 inhibitors that can interact with the BACE1-Arg235 side chain by a π-π stacking [43] as shown in **Figure 7B** (PDB ID: 4FSL). According to their structure–activity relationship study, the inhibitor possessing an electron-donating methoxy group on the *p*-position of phenyl ring that interacts with BACE1-Arg235 side chain can enhance BACE1-inhibitory activity. This finding indicated that an inhibitor possessing a P_2-aromatic ring with a higher electron density could strongly bind to the electron-poor π-orbital on the guanidino plane of the BACE1-Arg235 side chain. We thought that inhibitors formulated on the basis of such a quantum chemical interaction could never be designed using a classical concept on the basis of Newtonian mechanics, such as MM/MD calculation. Hence, we proposed the new "electron-donor bioisostere," concept, which involves quantum chemical interaction with an electron-poor π-orbital, such as the guanidine group of Arg235 [24].

We hypothesized that the quantum chemical interaction between an inhibitor and the side chain of BACE1-Arg235 plays a critical role in the inhibition mechanism. Therefore, we focused on the optimization around the P_2 region. The finding of a structure–activity relationship study focusing on the inhibitor's P_2 region is shown in **Table 1** [39–41, 44]. Inhibitors **21**, **22**, **25**, **27**, and **30** with hydrophobic and small-sized functional methanesulfonyl, methoxy, methylmercaptan, azide, and methyl groups on the P_2-pyridine ring display a higher inhibitory activity than inhibitors with a bulky or a hydrophilic group such as inhibitors **23**, **24**, **26**, **28**, and **29**. On the basis of the "electron-donor bioisostere" concept, we speculated that an electron-rich halogen atom could interact with the electron-poor guanidine π-orbital by Coulomb's force. Using the *ab initio* molecular orbital approach, Imai et al. described the slightly stronger calculated Cl-π interaction energy than the CH-π interaction and reported that its energy was affected by π-electron density [45]. Hence, we designed inhibitors **31–33** possessing a halogen atom on the P_2-pyridine ring. Inhibitors **31–36** exhibited more potent inhibitory activities (IC$_{50}$ values: 22, 15, and 24 nM, respectively). Next, inhibitors **34–36** possessing a fluorine atom on the *p*-position of P_3-phenyl group exhibited the potent inhibitory activities (IC$_{50}$ values: 13, 9, and 10 nM, respectively) [41]. Among them, inhibitor **35** (KMI-1303) exhibited the most potent inhibitory activity and is available from Wako Pure Chemical Industries (Japan) as a reagent for biological research.

4. Conclusion

Pyridines are important in medicinal chemistry because of their properties, which include weak basicity, water solubility, *in vivo*/chemical stability, hydrogen bond-forming ability, and small molecular size. Pyridine moieties are incorporated in many drugs and pesticides. Water solubility is one role of pyridines. The replacement of a portion of drugs with a pyridine moiety can improve their water solubility for the development of practical drugs that are suitable for an orally administrated or an injectable formulation. This approach is also applicable to the prodrug strategy. The bioisostere is one of the important roles of pyridines in drug design. The abovementioned attributes of pyridines enable the application as a bioisostere of amines, amides, heterocyclic rings containing a/some nitrogen atoms, and the benzene ring. The replacement of a part of lead compounds with pyridines is an important tool for the development of practical drugs. Moreover, the replacement of a portion of a drug molecule with pyridines might control the selectivity against subtypes of a target biomolecule, such as the histamine H_1 receptor antagonist, mepyramine. Recently, we reported a series of BACE1 inhibitors possessing a pyridine scaffold. In this process, we observed an important and third role of pyridines in medicinal chemistry, other than the conventional role of pyridines. This chapter has discussed on this third role. The replacement of the isophthalic scaffold of inhibitors with a pyridinedicarboxylic scaffold enables the control of the conformation of inhibitors. We designed the potent BACE1 inhibitor KMI-1303 using the design approach based on a conformer of the docked inhibitor in the target molecule—the *in silico* conformational structure-based design.

Acknowledgements

This study was supported in part by the Grants in Aid for Scientific Research from MEXT (Ministry of Education, Culture, Sports, Science and Technology), Japan (KAKENHI No. 23590137 and No. 26460163), and a donation from Professor Emeritus Tetsuro Fujita of Kyoto University. Prof. Fujita passed away on January 1 of last year. Prof. Fujita was my teacher. We dedicate this chapter to Prof. Fujita.

Conflict of interest

We confirmed independence from the funding source.

Author details

Yoshio Hamada

Address all correspondence to: pynden@gmail.com

Faculty of Frontier of Innovative Research in Science and Technology, Konan University, Kobe, Japan

References

[1] Altaf AA, Shahzad A, Gul Z, et al. A review on the medicinal importance of pyridine derivatives. Journal of Drug Design and Medicinal Chemistry. 2015;**1**:1. DOI: 10.11648/j.jddmc.20150101.11

[2] Felson DT, Smolen JS, Wells G, et al. American College of Rheumatology/European league against rheumatism provisional definition of remission in rheumatoid arthritis for clinical trials. Arthritis and Rheumatism. 2011;**63**:573. DOI: 10.1002/art.30129

[3] Ohwada J, Tsukazaki M, Hayase T, et al. Design, synthesis and antifungal activity of a novel water soluble prodrug of antifungal triazole. Bioorganic & Medicinal Chemistry Letters. 2003;**13**:191-196. DOI: 10.1016/S0960-894X(02)00892-2

[4] Priyanka LG, Priyanka SG, Deepali MJ, et al. The use of bioisosterism in drug design and molecular modification. American Journal of PharmTech Research. 2012;**2**:1-23. DOI: 10.1.1.301.520

[5] Parsons ME, Ganellin CR. Histamine and its receptors. British Journal of Pharmacology. 2006;**147**:S127-S135. DOI: 10.1038/sj.bjp.0706440

[6] Shimamura T, Shiroishi M, Weyand S, et al. Structure of the human histamine H_1 receptor complex with doxepin. Nature. 2011;**475**:65-70. DOI: 10.1038/nature10236

[7] Supuran CT, Winum J-Y, editors. Drug Design of Zinc-Enzyme Inhibitors: Functional, Structural, and Disease Applications. Wiley; 2009. pp. 487-672. DOI: 10.1002/9780470508169

[8] Kiyama R, Tamura Y, Watanabe F, et al. Homology modeling of gelatinase catalytic domains and docking simulations of novel sulfonamide inhibitors. Journal of Medicinal Chemistry. 1999;**42**:1723-1738. DOI: 10.1021/jm980514x

[9] Selkoe DJ. Toward a comprehensive theory for Alzheimer's disease. Hypothesis: Alzheimer's disease is caused by the cerebral accumulation and cytotoxicity of amyloid β-protein. Annals of the New York Academy of Sciences. 2000;**924**:17-25. DOI: 10.1111/j.1749-6632.2000.tb05554.x

[10] Selkoe DJ. The deposition of amyloid proteins in the aging mammalian brain: Implications for Alzheimer's disease. Annals of Medicine. 1989;**21**:73-76. DOI: 10.3109/07853898909149187

[11] Selkoe DJ. Translating cell biology into therapeutic advances in Alzheimer's disease. Nature. 1999;**399**:A23-A31. DOI: 10.1038/399a023

[12] Sinha S, Lieberburg I. Cellular mechanisms of β-amyloid production and secretion. Proceedings of the National Academy of Sciences of the United States of America. 1999;**96**:11049-11053. DOI: 10.1073/pnas.96.20.11049

[13] Vassar R, Bennett BD, Babu-Khan S, et al. β-Secretase cleavage of Alzheimer's amyloid precursor protein by the transmembrane aspartic protease BACE. Science. 1999;**286**:735-741. DOI: 10.1523/JNEUROSCI.3657-09.2009

[14] Hussain I, Powell D, Howlett DR, et al. Identification of a novel aspartic protease (asp 2) as β-Secretase. Neuroscience. 1999;**14**:419-427. DOI: 10.1523/JNEUROSCI.3657-09.2009

[15] Yan R, Bienkowski MJ, Shuck ME, et al. Membrane-anchored aspartyl protease with Alzheimer's disease β-secretase activity. Nature. 1999;**402**:533-537. DOI: 10.1038/990107

[16] Sinha S, Anderson JP, Barbour R, et al. Purification and cloning of amyloid precursor protein β-secretase from human brain. Nature. 1999;**402**:537-540. DOI: 10.1038/990114

[17] Turner PR, O'Connor K, Tate WP, Abraham WC. Roles of amyloid precursor protein and its fragment in regulating neural activity, plasticity and memory. Progress in Neurobiology. 2003;**70**:1-32. DOI: 10.1016/S0301-0082(03)00089-3

[18] Fluhrer R, Capell A, Westmeyer G, et al. A non-amyloidogenic function of BACE-2 in the secretory pathway. Journal of Neurochemistry. 2002;**81**:1011-1020. DOI: 10.1046/j. 1471-4159.2002.00908.x

[19] Roberds SL, Anderson J, Basi G, et al. BACE knockout mice are healthy despite lacking the primary β-secretase activity in brain: Implications for Alzheimer's disease therapeutics. Human Molecular Genetics. 2001;**10**:1317-1324. DOI: 10.1093/hmg/10.12.1317

[20] Hamada Y, Kiso Y. New directions for protease inhibitors directed drug discovery. Biopolymers. 2016;**106**:563-579. DOI: 10.1002/bip.22780

[21] Hamada Y, Kiso Y. Aspartic protease inhibitors as drug candidates for treating various difficult-to-treat diseases. In: Ryadnov M, Farkas E, editors. Amino Acids, Peptides and Proteins, Vol. 39. London: Royal Society of Chemistry; 2015. pp. 114-147. DOI: 10.1039/9781849739962-00114

[22] Hamada Y. Drug discovery of b-secretase inhibitors based on quantum chemical interactions for the treatment of Alzheimer's disease. SOJ Pharmacy & Pharmaceutical Sciences. 2014;**1**;1-8. DOI: 10.15226/2374-6866/1/3/00118

[23] Hamada Y, Kiso Y. Recent progress in the drug discovery of non-peptidic BACE1 inhibitors. Expert. Opinion on Drug Discovery. 2009;**4**:391-416. DOI: 10.1517/17460440902806377

[24] Hamada Y, Kiso Y. The application of bioisosteres in drug design for novel drug discovery: Focusing on acid protease inhibitors. Expert Opinion on Drug Discovery. 2012;**7**:903-922. DOI: 10.1517/17460441.2012.712513

[25] Hamada Y, Kiso Y. Advances in the identification of β-secretase inhibitors. Expert Opinion on Drug Discovery. 2013;**8**:709-731. DOI: 10.1517/17460441.2013.784267

[26] Nguyen J-T, Hamada Y, Kimura T, Kiso Y. Design of potent aspartic protease inhibitors to treat various diseases. Archiv der Pharmazie [Chemistry in Life Sciences]. 2008;**341**:523-535. DOI: 10.1002/ardp.200700267

[27] Ghosh AK, Shin D, Downs D, et al. Design of potent inhibitors for human brain memapsin 2 (β-secretase). Journal of American Chemistry Society. 2000;**122**:3522-3523. DOI: 10.1021/ja000300g

[28] Hong L, Koelsch G, Lin X, et al. Structure of the protease domain of memapsin 2 (β-secretase) complexed with inhibitor. Science. 2000;**290**:150-153. DOI: 10.1126/science.290.5489.150

[29] Ghosh AK, Bilcer G, Harwood C, et al. Structure-based design: Potent inhibitors of human brain memapsin 2 (β-secretase). Journal of Medicinal Chemistry. 2001;**44**:2865-2868. DOI: 10.1021/jm0101803

[30] Hong L, Turner RT, Koelsch G, et al. Crystal structure of memapsin 2 (β-secretase) in complex with an inhibitor OM00-3. Biochemistry. 2002;**41**:10963-10967. DOI: 10.1021/bi026232n

[31] Kimura T, Shuto D, Kasai K, et al. KMI-358 and KMI-370, highly potent and small-sized BACE1 inhibitors containing phenylnorstatine. Bioorganic & Medicinal Chemistry Letters. 2004;**14**:1527-1531. DOI: 10.1016/j.bmcl.2003.12.088

[32] Kimura T, Shuto D, Hamada Y, et al. Design and synthesis of highly active Alzheimer's β-secretase (BACE1) inhibitors, KMI-420 and KMI-429, with enhanced chemical stability. Bioorganic & Medicinal Chemistry Letters. 2005;**15**:211-215. DOI: 10.1016/j.bmcl.2004.09.090

[33] Asai M, Hattori C, Iwata N, et al. The novel β-secretase inhibitor KMI-429 reduces amyloid beta peptide production in amyloid precursor protein transgenic and wild-type mice. Journal of Neurochemistry. 2006;**96**:533-540. DOI: 10.1111/j.1471-4159.2005.03576.x

[34] Kimura T, Hamada Y, Stochaj M, et al. Design and synthesis of potent β-secretase (BACE1) inhibitors with P1' carboxylic acid bioisostere. Bioorganic & Medicinal Chemistry Letters. 2006;**16**:2380-2386. DOI: 10.1016/j.bmcl.2006.01.108

[35] Hamada Y, Igawa N, Ikari H, et al. β-Secretase inhibitors: Modification at the P_4 position and improvement of inhibitory activity in cultured cells. Bioorganic & Medicinal Chemistry Letters. 2006;**16**:4354-4359. DOI: 10.1016/j.bmcl.2006.05.046

[36] Hamada Y, Abdel-Rahman H, Yamani A, et al. BACE1 inhibitors: Optimization by replacing the P_1' residue with non-acidic moiety. Bioorganic & Medicinal Chemistry Letters. 2008;**18**:1649-1653. DOI: 10.1016/j.bmcl.2008.01.058

[37] Tagad HD, Hamada Y, Nguyen J-T, et al. Design of pentapeptidic BACE1 inhibitors with carboxylic acid bioisosteres at P_1' and P_4 positions. Bioorganic & Medicinal Chemistry. 2010;**18**:3175-3186. DOI: 10.1016/j.bmc.2010.03.032

[38] Tagad HD, Hamada Y, Nguyen J-T, et al. Structure-guided design and synthesis of P_1' position 1-phenylcycloalkylamine-derived pentapeptidic BACE1 inhibitors. Bioorganic & Medicinal Chemistry. 2011;**19**:5238-5246. DOI: 10.1016/j.bmc.2011.07.002

[39] Hamada Y, Ohta H, Miyamoto N, et al. Novel non-peptidic and small-sized BACE1 inhibitors. Bioorganic & Medicinal Chemistry Letters. 2008;**18**:1654-1658. DOI: 10.1016/j.bmcl.2008.01.056

[40] Hamada Y, Tagad HD, Nishimura Y, et al. Tripeptidic BACE1 inhibitors devised by in-silico conformational structure-based design. Bioorganic & Medicinal Chemistry Letters. 2014;**22**:1130-1135

[41] Hamada Y, Ohta H, Miyamoto N, et al. Significance of interaction of BACE1-Arg235 with its ligands and design of BACE1 inhibitors with P_2 pyridine scaffold. Bioorganic & Medicinal Chemistry Letters. 2009;**19**:2435-2439. DOI: 10.1016/j.bmcl.2009.03.049

[42] Crowley PB, Golovin A. Cation-π interactions in protein-protein interfaces. Proteins. 2005;**59**:231-239. DOI: 10.1002/prot.20417

[43] Gerritz SW, Zhai W, Shi S, et al. Acyl guanidine inhibitors of β-Secretase (BACE-1): Optimization of a micromolar hit to a nanomolar lead via iterative solid- and solution-phase library synthesis. Journal of Medicinal Chemistry. 2012;**55**:9208-9223. DOI: 10.1021/jm300931y

[44] Hamada Y, Suzuki K, Nakanishi T, et al. Structure-activity relationship study of BACE1 inhibitors possessing a chelidonic or 2,6-pyridinedicarboxylic scaffold at the P2 position. Bioorganic & Medicinal Chemistry Letters. 2014;**24**:618-623

[45] Imai YN, Inoue Y, Nakanishi I, Kitaura K. Cl-π interactions in protein–ligand complexes. Protein Science. 2008;**17**:1129-1137

Diverse Promotive Action of Pyridinecarboxylic Acids on Flowering in Ornamentals and Seedling Growth in Vegetable Crops

Shigeru Satoh

Additional information is available at the end of the chapter

http://dx.doi.org/10.5772/intechopen.75636

Abstract

This chapter describes our recent findings on diverse biological effects of pyridinecarboxylic acids, both pyridine-*di*-carboxylic acids (PDCAs) and pyridine-*mono*-carboxylic acids (PCAs), on plant growth processes. PDCA analogs promoted flowering and extended display time (vase life) of cut flowers of spray-type carnation. 2,3-PDCA and 2,4-PDCA were most active in the promotion. Apart from these actions, some of PDCAs and PCAs stimulated root and shoot growth of lettuce, carrot, and rice seedlings. Studies on structure–activity relationship of the chemicals showed that one of the most effective chemicals was pyridine-3-carboxylic acid. Pyridine-3-carboxylic acid is known as vitamin B3 (niacin) and safe for human and animals. These findings suggested the possibility to develop PDCAs and PCAs as novel flower-care agents as well as growth-promoting agents which will be used for vegetable cultivation.

Keywords: pyridine-*di*-carboxylic acids, pyridine-*mono*-carboxylic acids, pyridine-3-carboxylic acid, flowering, seedling growth, promotion

1. Introduction

Pyridine-di-carboxylic acid (PDCA) has six structural analogs, that is, 2,3-, 2,4-, 2,5-, 2,6-, 3,4-, and 3,5-PDCA. In the past, there were a few works on biological activities of PDCA analogs. 2,6-PDCA was reported to inactivate markedly aconitase, but 2,3-, 2,4-, and 2,5-PDCA did not [1]. 2,4-PDCA is a structural analog of 2-oxoglutarate (OxoGA) and was shown to inhibit OxoGA-dependent dioxygenases by competing with OxoGA [2, 3]. For example, OxoGA-dependent dioxygenases include proline-4-hydroxylase [2–4] and enzymes involved in gibberellin (GA) biosynthesis and metabolism, such as gibberellin 3β-dioxygenase (gibberellin 3β-hydroxylase), gibberellin-44 dioxygenase, and gibberellin 2β-dioxygenase (gibberellin 2β-hydroxylase) [5–9].

Iturriagagoittia-Bueno et al. [10] reported that OxoGA competitively inhibited the activity of 1-aminocyclopropane-1-carboxylate (ACC) oxidase with respect to ascorbate, which was extracted from ripe pear fruits. ACC oxidase catalyzes the last-step reaction of ethylene bio-synthesis in plants. Vlad et al. [4] demonstrated that 2,4-PDCA inhibited ethylene produc-tion in detached carnation flowers and delayed senescence of the flowers and suggested that 2,4-PDCA inhibited the activity of ACC oxidase by acting as a structural analog of ascorbate. Then, Fragkostefanakis et al. [11] showed that 2,4-PDCA inhibited the in vitro activity of ACC oxidase prepared from tomato pericarp tissues. These results confirmed that 2,4-PDCA inhib-its ACC oxidase by competing with ascorbate.

Recently, we examined 2,4-PDCA action on ethylene production and senescence in spray-type carnation flowers to know whether it can be used as a flower-care agent to prolong the display time of the flowers, which had not been referred in the previous paper [4]. Interestingly, we could demonstrate the acceleration of flower opening by 2,4-PDCA in cut spray-type carnation flowers, in addition to retardation of their senescence, resulting in prolonging their display time. We furthermore compared 2,4-PDCA action with those of its structural analogs, such as 2,3-PDCA, 2,5-PDCA, and so on, in order to obtain further information on 2,4-PDCA's action. On the other hand, in the course of the above-mentioned study, we found that 2,3-PDCA promoted root elongation, whereas 2,4-PDCA inhibited it in lettuce, carrot, and rice seedlings. We explored the promoting activities of 2,3-PDCA to other PDCA analogs and pyridine-mono-carboxylic acid (PCA) analogs. Also, we carried out a preliminary investigation on the possible biochemical and molecular mechanism of PDCA, mainly with 2,4-PDCA. This chapter describes the details of the effects of 2,4-PDCA and related chemicals on flower opening and display time in carna-tion flowers, as well as the promotion of seedling growth in some agricultural crops.

2. 2,4-PDCA extends the display time of cut spray-type carnation flowers by promoting flower opening as well as retarding senescence

2.1. Procedures for observation of flower opening and senescence

A carnation cultivar, *Dianthus caryophyllus* L. "Light Pink Barbara (LPB)," which belongs to the spray type of carnation flowers having multiple flowers (buds) on a stem, was used. Flowers were harvested when the first floret out of six to eight flower buds on a stalk was partially open. Stems of cut flowers were trimmed to 60 cm. Bunches of 5 stems, each having 5 flower buds (25 buds in total per bunch), were put in 900-ml glass jars containing 300 ml of test solutions. The test solutions were distilled water (control) and the solutions containing each of PDCA analogs at given concentrations. The flowers were kept under continuous light at 23°C and 50–70% relative humidity. Fully open non-senescent (FONS) flowers, which were regarded as flowers ranging from Open stage 6 to Senescence stage 2 [12, 13], were counted daily, and the percentage of these flowers to the total number (25) of initial flower buds per bunch was calculated. The display time of the cut flowers in days is expressed by the number of days during which the percentage of FONS flowers was 40% or more [14, 15]. Data are presented as changes of the percentages of FONS flowers during 24 days. The time to flower opening was defined as the number of days from the start of experiment to the time when the percentage of FONS flowers reached 40% [16].

2.2. Analysis of flower opening as well as display time (vase life) in cut "LPB" carnation flowers

Figure 1 shows changes in the percentage of FONS flowers for cut "LPB" carnation flowers treated with different concentrations of 2,4-PDCA [16]. The treatment with 2,4-PDCA tended to shorten the time to flower opening, which was 4.4 days for the control, 4.3 days for 0.3 mM, 3.3 days for 1 mM, and 3.8 days for 2 mM, although it was significantly different from the control only with treatment at 1 mM 2,4-PDCA by Steel's multiple range test ($P < 0.05$). These observations suggested that 2,4-PDCA has an activity to accelerate flower bud opening. The display time was significantly lengthened by treatment with 2,4-PDCA, attaining 53, 111, and 135% increases at 0.3, 1, and 2 mM 2,4-PDCA, respectively, as compared with the control [15].

This figure was adapted from **Figure 2** in Sugiyama and Satoh [16]. The percentage of fully open and non-senescent (FONS) flowers was calculated from the proportion of those flowers to the total number of initial flower buds (25 buds per 5 flowers). The time to flower opening (I) is the time in days from the start of experiment until the percentage of open flowers reached at 40%. The display time of the flowers (II) is the duration when the percentage of FONS flowers was 40% or more.

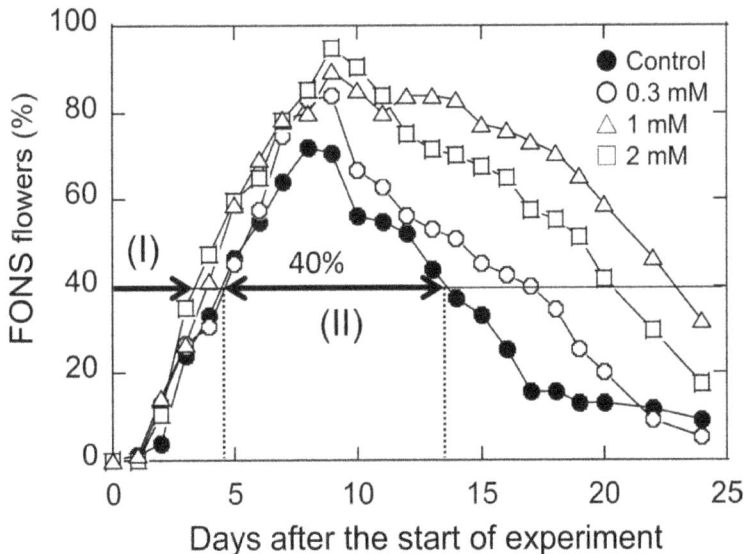

Figure 1. Profiles of flower opening and duration of display time in cut flowers of 'Light Pink Barbara' (LPB) carnation treated continuously with different concentrations of 2,4-PDCA. This figure was adapted from **Figure 2** in Sugiyama and Satoh [16]. The percentage of fully-open and non-senescent (FONS) flowers was calculated from the proportion of those flowers to the total number of initial flower buds (25 buds per 5 flowers). The time to flower opening (I) is the time in days from the start of experiment until the percentage of open flowers reached 40%. The display time of the flowers (II) is the duration when the percentage of FONS flowers was 40% or more.

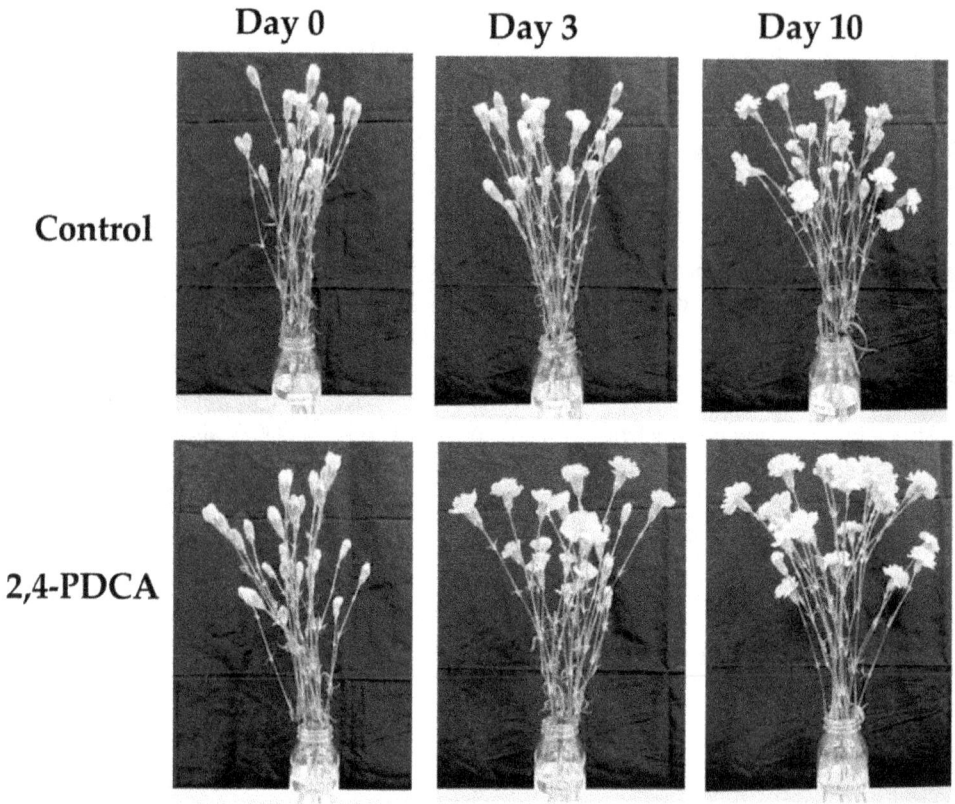

Figure 2. Flower opening and senescing profiles of cut "LPB" carnation flowers treated without (control) or with 2,4-PDCA at 2 mM. Typical profiles for each treatment out of three replicates were chosen for 0, 3, and 10 days of the experiment. This figure was adapted from **Figure 4** in Sugiyama and Satoh [16], by extracting only the data for 2,4-PDCA.

2.3. Comparison of the effect of 2,4-PDCA between the continuous and pulse treatments

In the experiment described earlier, 2,4-PDCA was applied to cut carnation flowers continuously during experiments (the continuous treatment). This procedure sometimes caused detrimental side effects, resulting in the browning of leaves or broken stalks [16], probably because of excess absorption of the chemicals. Also, this procedure seems to be practically inadequate from the application perspective of the chemicals, since it would need much labor work. Therefore, we tried to apply PDCA by a pulse treatment, in which the flowers were treated once after harvest for a short period when kept in water [17]. We treated cut "LPB" flowers with 2,4-PDCA at 0 (control), 5, or 10 mM for 24 h, and thereafter left with their stalk end in water (pulse treatment). The time to flower opening was 6.0 ± 0.5 days (shown by the mean $\pm$ SE of three replicates) in the control. The pulse treatment with 5 and 10 mM 2,4-PDCA shortened it to 5.0 ± 0.7 days and 3.1 ± 0.4 days, respectively, whereas the continuous treatment with 2 mM 2,4-PDCA shortened the time to flower opening to 4.3 ± 0.3 days. The display time in the control was 5.7 ± 1.7 days. It was lengthened by the pulse treatment with 2,4-PDCA at 5 mM to 11.9 ± 0.7 days and with 10 mM to 14.2 ± 0.3 days, and by the continuous treatment with 2 mM 2,4-PDCA to 11.2 ± 0.5 days. These results revealed that the effect of pulse treatment with 5 or 10 mM 2,4-PDCA on the flower opening characteristics was similar to or greater than that of the continuous treatment with 2 mM 2,4-PDCA in "LPB" carnation.

3. PDCA analogs extend the display time of carnation flowers

As described earlier, 2,4-PDCA stimulated flower opening and prolonged the display time in cut spray-type carnation flowers. Then, we compared 2,4-PDCA's action with those of its structural analogs (2,3-, 2,5-, 2,6-, 3,4-, and 3,5-PDCA) in order to obtain further information on 2,4-PDCA's action [16]. In this experiment, cut "LPB" carnation flowers were treated without (control) or with each of PDCA analogs at 2 mM.

Figure 2 shows flower opening and senescing profiles of the control and 2,4-PDCA-treated flowers, which were chosen as typical specimens, 0, 3, and 10 days after the start of the experiment. On day 3, open pink flowers were seen more among the flowers treated with 2,4-PDCA than in the control. These observations indicated that treatment with 2,4-PDCA accelerated flower opening as compared with the untreated control flowers. On day 10, the control flowers remained as a mixture of buds, open flowers, and senesced flowers. The senesced flowers showed in-rolling and wilting of petals, typical symptoms of senescence in response to

Chemicals	Time to flower opening (days)	Display time (days)
Control	9.0 ± 1.0	8.3 ± 1.9
2,3-PDCA	4.0*± 0.0	14.7*± 0.3
2,4-PDCA	4.0*± 0.6	14.7*± 1.7
2,5-PDCA	4.7*± 0.3	13.0 ± 2.1
2,6-PDCA	5.3*± 0.3	9.3 ± 1.5
3,4-PDCA	5.7*± 0.3	14.7*± 2.4
3,5-PDCA	4.7*± 0.7	15.5*± 1.2

Table 1. Effects of PDCA analogs on the time to flower opening and the display time of cut spray "LPB" carnation flowers. The data were adapted from **Table 1** in Sugiyama and Satoh [16], by extracting only the data for 2 mM PDCA. Data are shown in days with the mean ± SE of three replicates, each with five flower stems with five florets (buds) on a stem.* shows a significant difference from the control in each column by Dunnett's multiple range test (P < 0.05).

ethylene. On the other hand, almost all the flowers treated with 2,4-PDCA were fully open with non-wilted and turgid petals. At the later stage, 2,4-PDCA-treated flowers withered with browning at the petal margins, as well as with jumbled, but turgid, fading petals [15].

Table 1 summarizes the effects of all the PDCA analogs on the time to flower opening and the display time of cut "LPB" flowers. The time to flower opening was 9.0 days for the control flowers and was significantly shortened by treatment with all of the PDCA analogs. It was the shortest by treatment with 2,3-PDCA and 2,4-PDCA (both 4.0 days), followed by 2,5-PDCA and 3,5-PDCA (both 4.7 days) and 2,6-PDCA (5.3 days) and 3,4-PDCA (5.7 days). Four of the PDCA analogs (2,3-, 2,4-, 3,4-, and 3,5-PDCA) significantly lengthened the display time compared with that of the control (8.3 days); the display times varied from 14.7 to 15.5 days. 2,5-PDCA and 2,6-PDCA tended to lengthen the display time, although their effects were not statistically significant.

4. PDCA and PCA analogs stimulate root and shoot growth in lettuce, carrot, and rice seedlings

4.1. Effects of 2,3-PDCA and 2,4-PDCA on root elongation of lettuce, carrot, and rice seedlings

The action mechanism of PDCAs for accelerating flower bud opening in carnation remains unresolved. Satoh et al. [15] hypothesized on the association of gibberellin (GA) with the promoting action of 2,4-PDCA on flower bud opening in cut flowers of spray-type carnation. This hypothesis arose from the notion that 2,4-PDCA is a structural analog of OxoGA, which is a cosubstrate for enzymes acting in GA biosynthesis and inactivation. We tried to test whether 2,3-PDCA and 2,4-PDCA have GA-like activity using a bioassay system, in which exogenously applied GA promotes hypocotyl elongation of lettuce seedlings [18].

Uniformly germinated seeds were placed on solidified Gellan Gum (1.0%, w/v) in a glass test tube. The Gellan gel contained 2,3-PDCA or 2,4-PDCA at 0 (control), 0.3, 1, and 3 mM, and GA_3 at 0.3 mM. After 7 days at 23°C in the light, 2,3-PDCA at 0.1 mM promoted root elongation, and the degree of promotion increased up to 1 mM, then declined slightly at 3 mM, whereas, 2,4-PDCA at 0.1–3 mM severely inhibited the elongation of lettuce roots (**Figure 3**). GA_3 at 0.3 mM promoted hypocotyl elongation but inhibited root elongation in lettuce seedlings. Moreover, we found that 2,3-PDCA and 2,4-PDCA had similar effects on root elongation of carrot seedlings; 2,3-PDCA at 0.3–3 mM promoted the elongation of roots, whereas 2,4-PDCA at 0.1–3 mM inhibited it.

4.2. Effects of PDCA and PCA analogs on root and shoot elongation of rice seedlings

Similar to the investigation with lettuce and carrot seedlings, the effects of PDCA analogs on root growth of rice seedlings were examined. 2,3-, 3,4-, and 3,5-PDCA promoted the root elongation, although the latter two were less effective, whereas 2,4-PDCA and 2,6-PDCA inhibited the root elongation, and the effect of 2,5-PDCA on root elongation was very small. These

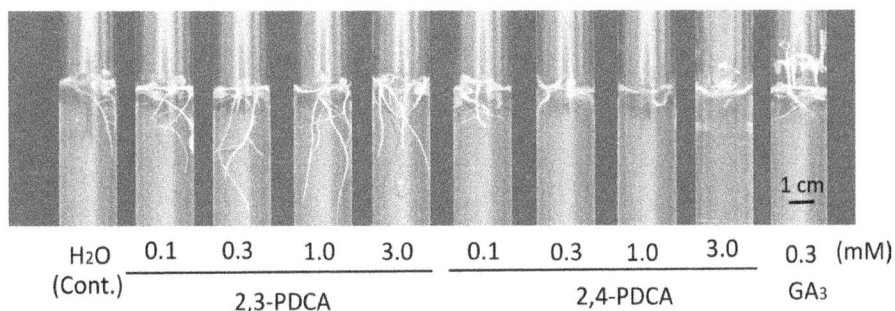

Figure 3. Effects of 2,3-PDCA, 2,4-PDCA, and GA$_3$ in root elongation of lettuce seedlings. This figure was cited from **Figure 1** in Satoh and Nomura [18]. Three germinated lettuce seeds with radicles protruded 1 mm were placed on the surface of solidified Gellan Gum (0.1%, w/v) containing 2,3-PDCA and 2,4-PDCA at 0.1–3 mM and GA$_3$ at 0.3 mM and allowed to grow for 7 days at 23°C under light from white fluorescent lamps. Water was used as the control.

results suggested that the carboxyl group at position 3 of the pyridine ring is necessary to pro-mote root elongation in rice seedlings and probably in lettuce and carrot seedlings. Therefore, we explored the activity of PCA analogs, that is, 2-, 3-, and 4-PCA and 3-PCA amide, as well as 2,3-, 2,4-, and 3,4-PDCA, on the growth of rice seedlings grown for 7 days by hydroponic culture [18].

Uniformly germinated rice seeds were hydroponically grown in water (control) and test solu-tions containing PDCA or PCA analogs (2,3-, 2,4-, and 3,4-PDCA; 2-, 3-, and 4-PCAs; 3-PCA amide) at 0.03, 0.1, and 0.3 mM. Twenty-five germinated seeds were floated on a plastic mesh using a polyurethane float in a transparent plastic box containing 300 ml of test solutions. After 7 days, the whole seedlings, and roots and shoots were photographed. The root and shoot (leaf + leaf sheath) lengths were measured with a curvemeter (Pen-type Map-meter Concurve 10, Koizumi Sokki Mfg. Co., Ltd., Nagaoka, Japan) on printed photographs. The total root length of each seedling was obtained as the sum of the lengths of a seminal root and all coronal roots. The seedlings were aligned according to the total root length, then 15 seedlings in the middle were chosen for measuring the root length. Also, the shoot length was determined similarly.

Figure 4 shows the seedlings treated with the PDCAs and PCAs at 0.3 mM. The apparent mass of roots (both seminal and coronal roots) was clearly increased by 3-PCA, 3,4-PDCA, and 2,3-PDCA, but decreased by 2-PCA, 4-PCA, and 2,4-PDCA. Interestingly, 3-PCA amide decreased the mass of roots, although not as much as 2-PCA, 4-PCA, and 2,4-PDCA. On the other hand, all the chemicals, except 4-PCA, appeared to promote the elongation of shoots, which was judged from the protrusion of shoot over the edge of containers.

Figure 5A and **B** show the total length of root and shoot per seedling, respectively, treated with PCA and PDCA analogs at 0.03–0.3 mM. Root elongation was significantly promoted by 3-PCA, 3,4-PDCA, and 2,3-PDCA at all the concentrations used. The magnitude of promotion was similar in 3-PCA and 3,4-PDCA, followed by 2,3-PDCA. By contrast, 2-PCA, 4-PCA, 2,4-PDCA, and 3-PCA amide significantly inhibited the root elongation at all the concentrations

Figure 4. Effects of PCA analogs (2-, 3-, 4-PCA, and 3-PCA amide) and PDCA analogs (2,3-, 2,4-, and 3,4-PDCA) on the growth of rice seedlings. Germinated rice seeds were grown at 23°C for 7 days hydroponically in test solutions containing each chemical at 0.3 mM. This figure was previously shown as **Figure 5** in Satoh and Nomura [18].

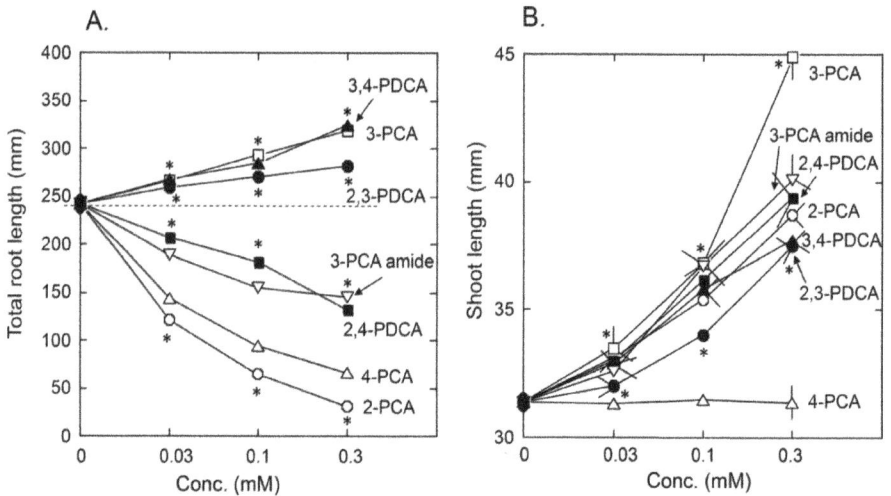

Figure 5. Effects of PCA analogs (2-, 3-, 4-PCA, and 3-PCA amide) and PDCA analogs (2,3-, 2,4-, and 3,4-PDCA) on the root elongation (A) and shoot elongation (B) of rice seedlings. * shows significant difference from the control by Dunnett's multiple range test ($P < 0.05$), and symbols between top and bottom stars at given concentrations are all significantly different from the control. This figure was previously shown as **Figure 6** in Satoh and Nomura [18].

used. 2-PCA had the greatest inhibitory effect on root elongation, followed by 4-PCA, 2,4-PDCA, and 3-PCA amide. On the other hand, all the chemicals, except 4-PCA, significantly promoted the elongation of shoots of rice seedlings. 3-PCA was most active in the promotion of shoot elongation in rice seedlings, followed by 3-PCA amide, 2,4-PDCA, and 2-PCA. 4-PCA did not affect the shoot elongation of rice seedlings.

Precise analyses of the effects of PDCA and PCA analogs revealed that the acceleration of root elongation in rice seedlings depends on the free carboxyl group (-COOH) at position 3 of the pyridine ring, and the inhibition of root elongation depends on -COOH at positions 2 or 4 of the ring (**Figures 3–6**). When the second carboxyl group was introduced into 3-PCA, its introduction to position 2 of the pyridine ring, which makes 2,3-PDCA, reduced the promoting activity of 3-PCA more severely than that into position 4 of the pyridine ring, which makes 3,4-PDCA (**Figures 4 and 5**). Moreover, in PDCAs, 3-COOH overcame the inhibitory effects of 2- or 4-COOH; in other words, the latter two -COOHs could not nullify the promotive effect of 3-COOH. Interestingly, when 3-COOH of 3-PCA was replaced with -CONH$_2$, the resultant 3-PCA amide lost the root elongation promoting activity in rice seedlings (**Figure 5**). Both 3-PCA and 3-PCA amide are vitamin B3, known as nicotinic acid and nicotinamide, respectively, and are regarded to have an activity equivalent to the vitamin. The present findings that 3-PCA promoted but 3-PCA amide inhibited rice root elongation suggested that the root elongation promoting the activity of 3-PCA did not originate from its activity as vitamin B3. The present findings showed the promoting effects of 3-PCA and PDCA analogs with 3-COOH on root elongation in rice seedlings. This notion may also apply to the promotion of root elongation in lettuce and carrot seedlings.

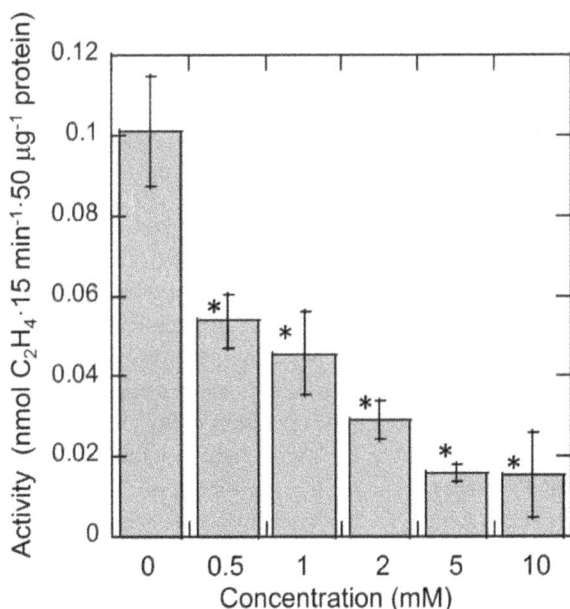

Figure 6. The inhibition of ACC oxidase activity by 2,4-PDCA. ACC oxidase was prepared by the expression of carnation *DcACO1* cDNA in *E. coli*, and its activity was determined at 1 mM ACC in the absence or presence of 2,4-PDCA at the given concentrations. Data are the means ± SE of triplicate determinations. This figure was adapted from **Figure 1** in Satoh et al. [15].

5. Search for a possible biochemical action mechanism of PDCA and PCA analogs

We searched for possible biochemical and molecular mechanism of PDCA action. So far, we tested effects of 2,4-PDCA on recombinant ACC oxidase activity [15] and on gibberellin action by examining the changes in gene expression of DELLA protein (GAI), a negative regulator of GA signaling, and GA contents in carnation flowers [19].

First, we examined the action of 2,4-PDCA on ACC oxidase, which was synthesized in *Escherichia coli* cells from carnation ACC oxidase gene (*DcACO1*). For the construction of an expression plasmid, the entire coding region was amplified from *DcACO1* cDNA [20] by PCR. DcACO1 protein was expressed from *DcACO1* cDNA in *E. coli* cells as described elsewhere [21]. ACC oxidase activity was assayed by the standard method [21]. The recombinant ACC oxidase had a Km of 118 mM for ACC. This Km value was comparable to that reported previously for ACC oxidase from carnation petals, 30–425 mM depending on CO_2 concentration [22] and 111–125 mM in the presence of $NaHCO_3$ [23].

Figure 6 shows the inhibitory effects of 2,4-PDCA on the activity of the recombinant ACC oxidase. 2,4-PDCA inhibited an enzyme activity by 47% at 0.5 mM, and the magnitude of inhibition increased gradually as its concentration increased, attaining 85% inhibition at 10 mM. The results confirmed that 2,4-PDCA inhibited the activity of ACC oxidase. Iturriagagoittia-Bueno et al. [10] showed that OxoGA inhibited ACC oxidase by competing with ascorbate. 2,4-PDCA was shown to compete with OxoGA in the inhibition of vertebrate collagen proline-4-hydroxylase, an OxoGA-dependent dioxygenase [2, 3].

Therefore, it is speculated that 2,4-PDCA inhibits ACC oxidase activity by competing with ascorbate, although there is a possibility that 2,4-PDCA inhibited ACC oxidase activity by chelating Fe^{2+} [24].

Next, we examined the expression profiles of *DcGAI* gene in carnation flowers by qRT-PCR analysis. *DcGAI* was expressed in all the tissues examined, that is, calyx, style, receptacle, ovary, and petals. A high level of expression was observed in calyx, and the expression levels in style, receptacle, ovary, and petals were low. We also analyzed the changes in expression levels of *DcGAI* in petals of carnation flowers during the flower opening process. The transcript level of *DcGAI* was high at the early stages of flower opening and decreased at later stages of flower opening. We examined the expression of the gene in 2,4-PDCA-treated flowers. Flowers at Opening stage (Os) 1 to Os 2 were treated with 2 mM 2,4-PDCA for 10 days. Using petals sampled from these flowers, the expression of *DcGAI* was examined by qRT-PCR. The expression of the gene was decreased in accordance with the progression of the flower opening process in the control flowers. By contrast, the expression of *DcGAI* was maintained at a high level in 2,4-PDCA-treated flowers and was significantly higher than that in the control flowers at day 10.

Thus, an increase in the transcript level of *DcGAI* was observed in 2,4-PDCA-treated flowers. GAI is a negative regulator of GA action and suppresses growth and elongation [25]. Therefore, it is supposed that the increase in *DcGAI* expression leads to the suppression of petal cell growth, which does not coincide with the enhancement of flower opening. Therefore, it is likely that the alteration of *GAI* expression is not related to the enhancing effect of 2,4-PDCA.

Furthermore, we examined GA contents as affected by 2,4-PDCA treatment in opening carnation flowers by LC–MS/MS analysis. Although endogenous GAs acting in carnation have not been identified so far, we set out to measure the GA_1 content since several ornamental plants (*Chrysanthemum morifolium, Eustoma grandiflorum, Gentiana triflora, Phalaenopsis hybrid*, and petunia) contain GA_1 as a bioactive GA [26]. However, GA_1 was not detected in carnation petals in our LC–MS/MS analysis. We found GA_3 accumulated instead and determined the GA_3 content in petals of opening flowers. In this experiment, all the flowers starting from Os 2 reached Os 4–6 in 4 days in the control flowers. 2,4-PDCA treatment accelerated flower opening, and all the treated flowers reached Os 4–6 in 2 days. We measured GA_3 content in the non-treated flowers at days 0–4 and found that the GA_3 level tended to be decreased in the course of flower opening. We also measured GA_3 content in 2,4-PDCA-treated flowers at day 1, when the treated flowers showed a significant increase in the number of open flowers. We observed a tendency that GA_3 content in the 2,4-PDCA-treated flowers was lower than that in the control. The GA_3 content in the control flowers was 48.5 ± 10.0 pmol·g^{-1}FW, whereas it was 26.6 ± 14.3 pmol·g^{-1}FW in the 2,4-PDCA-treated flowers. There was no significant difference between the control and the treated samples by t-test at $P < 0.05$. These results showed that 2,4-PDCA increases the gene expression of the growth suppressor, GAI, and decreases the GA level, suggesting that GA signaling and action are altered by 2,4-PDCA treatment. However, such changes are contradictory to the enhancement of flower opening, which suggests that GA is not associated with the enhancing effect of 2,4-PDCA in carnation flowers.

6. Discussion

Our studies showed that PDCAs accelerate flower opening and retard senescence, which increase the number of open flowers, resulting in the extension of the display time of cut flowers of "LPB" carnation. At first, this finding was obtained by the experiment with 2,4-PDCA and then subsequent experiments with other PDCA analogs, that is, 2,3-, 2,5-, 2,6-, 3,4-, and 3,5-PDCA. Judging from their effectiveness in the acceleration of flower opening and extension of display time, 2,3-PDCA and 2,4-PDCA were thought to be suitable agents for the treatment of the flowers.

2,4-PDCA is a structural analog of OxoGA and has been suggested to inhibit OxoGA-dependent dioxygenases by competing with OxoGA and ethylene production in detached carnation flowers by competing with ascorbate on ACC oxidase action [4, 11]. We confirmed that 2,4-PDCA inhibited ACC oxidase action with a recombinant enzyme produced in *E. coli* cells from carnation *DcACO1* cDNA. These findings suggested that 2,4-PDCA primarily inhibited ACC oxidase action, resulting in the inhibition of ethylene production and the delay of withering of carnation flowers. However, probably, this mechanism of action could not be applicable for PDCA analogs other than 2,4-PDCA. Meanwhile, it was suggested that GA might be involved in the action of 2,4-PDCA on flower opening and senescence of carnation flowers [15]. However, 2,4-PDCA treatment gave no significant effects on the expression of *DcGAI* gene, which is a key factor in the GA-signaling pathway and on GA content in the flower petals

[19]. Therefore, at present, there is no reliable explanation for the PDCAs' action mechanism in the stimulation of flower opening and extension of display time in carnation flowers.

It was revealed that 2,3-PDCA promoted root elongation in lettuce, carrot, and rice seedlings, whereas 2,4-PDCA inhibited it [18]. The action of PDCA and PCA analogs on root and shoot elongation was further examined using rice seedlings. 2,3-, 3,4-, and 3,5-PDCA promoted rice root elongation, whereas 2,4- and 2,6-PDCA inhibited it, and 2,5-PDCA had little effect. 3-PCA (nicotinic acid) promoted rice root elongation, but 2- and 4-PCA did not. Moreover, 3-PCA amide (nicotinamide) did not promote root elongation. These findings indicated that a carboxyl group substituted on position 3 of the pyridine ring is necessary to promote root elongation, and that the promoting effect of 3-PCA was not from its action as vitamin B3, but from its intrinsic activity as a 3-COOH-substituted pyridine. On the other hand, all the PCA and PDCA analogs tested in this study, except for 4-PCA, promoted shoot elongation of rice seedlings. The mechanism of action of PCA and PDCA analogs on root growth will hopefully be elucidated in the near future.

We have observed that PDCAs are practically applicable to cut flowers of spray-type carnation cultivars other than "LPB." In addition, this issue will also be examined with other species of ornamentals, of which flowers are used as spray-type flowers, such as *Eustoma*, *Gypsophila*, and *Alstroemeria* flowers and spray-type *Chrysanthemum*. The promotion of root elongation in the seedling of vegetable and ornamental crops by PDCA and PCA analogs probably has merits in practical agriculture, since massive roots of seedlings surely promote root establishment of the seedlings after transplanting. From this point of view, we are now investigating a practical method to apply these chemicals to grow sound and vigorous seedlings of vegetable and ornamental crops.

Acknowledgements

This study was supported financially by a Matching Planner Program Grant (MP27115663028 to S. Satoh) from the Japan Science and Technology Agency and a Grant-in-Aid (16 K07604 to S. Satoh) for Scientific Research from the Japan Society for the Promotion of Science.

Conflict of interest

The author declares no conflict of interest.

Author details

Shigeru Satoh

Address all correspondence to: ssatoh@agr.ryukoku.ac.jp

Faculty of Agriculture, Ryukoku University, Otsu, Japan

References

[1] Murakami K, Haneda M, Hosokawa Y, Yoshino M. Prooxidant action of pyridine carboxylic acids: Transition metal-dependent generation of reactive oxygen species. Trace Nutrition Research. 2007;**24**:49-55. (In Japanese with English summary)

[2] Kivirikko KI, Myllyharju J. Prolyl 4-hydroxylases and their protein disulfide isomerase subunit. Matrix Biology. 1998;**6**:357-368

[3] Kivirikko KI, Pihlajaniemi T. Collagen hydroxylases and the protein disulfide isomerase subunit of prolyl 4-hydroxylases. Advances in Enzymology and Related Areas of Molecular Biology. 1998;**72**:325-398

[4] Vlad F, Tiainen P, Owen C, Spano T, Daher FB, Oualid F, Senol NO, Vlad D, Myllyharju J, Kalaitzis P. Characterization of two carnation petal prolyl 4 hydroxylases. Physiologia Plantarum. 2010;**140**:199-207

[5] Hedden P, Kamiya Y. Gibberellin biosynthesis: Enzymes, genes and their regulation. Annual Review of Plant Physiolology and Plant Molecular Biology. 1997;**48**:431-460

[6] Lange T, Hedden P, Graebe JE. Expression cloning of a gibberellin 20-oxidase, a multifunctional enzyme involved in gibberellin biosynthesis. Procedings of National Academy of Sciences USA. 1994a;**91**:8552-8556

[7] Lange T, Schweimer A, Ward DA, Hedden P, Graebe JE. Separation and characterization of three 2-oxoglutarate-dependent dioxygenases from *Cucurbita maxima* L. endosperm involved in gibberellin biosynthesis. Planta. 1994b;**195**:98-107

[8] Smith VA, MacMillan J. Purification and partial characterization of a gibberellin 2β-hydroxylase from *Phaseolus vulgaris*. Journal of Plant Growth Regulation. 1984;**2**:251-264

[9] Smith VA, MacMillan J. The partial purification and characterisation of gibberellin 2β-hydroxylases from seeds of *Pisum sativum*. Planta. 1986;**167**:9-19

[10] Iturriagagoittia-Bueno T, Gibson EJ, Schofield CJ, John P. Inhibition of 1-aminocyclopropane-1-carboxylate oxidase by 2-oxoacid. Phytochemistry. 1996;**43**:343-349

[11] Fragkostefanakis S, Kalaitzis P, Siomos AS, Gerasopoulos D. Pyridine 2,4-dicarboxylate downregulated ethylene production in response to mechanical wounding in excised mature green tomato pericarp discs. Journal of Plant Growth Regulation. 2013;**32**:140-147

[12] Harada T, Torii Y, Morita S, Masumura T, Satoh S. Differential expression of genes identified by suppression subtractive hybridization in petals of opening carnation flowers. Journal of Experimental Botany. 2010;**61**:2345-2354

[13] Morita S, Torii Y, Harada T, Kawarada M, Onodera R, Satoh S. Cloning and characterization of a cDNA encoding sucrose synthase associated with flower opening through early senescence in carnation (*Dianthus caryophyllus* L.). Journal of Japanese Society for Horticultural Science. 2011;**80**:358-364

[14] Satoh S, Nukui H, Inokuma T. A method for determining the vase life of cut spray carnation flowers. Journal of Applied Horticulture. 2005;7:8-10

[15] Satoh S, Kosugi Y, Sugiyama S, Ohira I. 2,4-Pyridinedicarboxylic acid prolongs the vase life of cut flowers of spray carnations. Journal of Japanese Society for Horticultural Science. 2014;83:72-80

[16] Sugiyama S, Satoh S. Pyridinedicarboxylic acids prolong the vase life of cut flowers of spray-type 'light pink Barbara' carnation by accelerating flower opening in addition to an already-known action of retarding senescence. Horticulture Journal. 2015;84:172-177

[17] Satoh S, Nomura Y, Morita S, Sugiyama S. Further characterization of the action of pyridinedicarboxylic acids: Multifunctional flower care agents for cut flowers of spray-type carnation. Journal of Applied Horticulture. 2016;18:3-6

[18] Satoh S, Nomura Y. Promotion of root elongation by pyridinecarboxylic acids known as novel cut flower care agents. Plant Root. 2017;11:40-47

[19] Morita S, Sugiyama S, Nomura Y, Masumura T, Satoh S. Gibberellin is not associated with the enhancing effect of 2,4-pyridinedicarboxylic acid on flower opening of 'Light Pink Barbara' carnation. Horticulture Journal. 2017;86:519-527

[20] Kosugi Y, Shibuya K, Tsuruno N, Iwazaki Y, Mochizuki A, Yoshioka T, Hashiba T, Satoh S. Expression of genes responsible for ethylene production and wilting are differently regulated in carnation (Dianthus caryophyllus L.) petals. Plant Science. 2000;158:139-145

[21] Satoh S, Kosugi Y. Escherichia coli-based expression and in vitro activity assay of 1-aminocyclopropane-1-carboxylate (ACC) synthase and ACC oxidase. In: Binder BM, Schaller GM, editors. Ethylene Signaling: Methods and Protocols. New York: Springer; 2017. pp. 47-58

[22] Nijenhuis-de Vries MA, Woltering EJ, De Vrije T. Partial characterization of carnation petal 1-aminocyclopropane-1-carboxylate oxidase. Journal of Plant Physiology. 1994;144: 549-554

[23] Kosugi Y, Oyamada N, Satoh S, Yoshioka T, Onodera E, Yamada Y. Inhibition by 1-aminocyclobutane-1-carboxylate of the activity of 1-aminocyclopropane-1-carboxylate oxidase obtained from senescing petals of carnation (Dianthus caryophyllus L.) flowers. Plant & Cell Physiology. 1997;38:312-318

[24] Smith JJ, Ververidis P, John P. Characterization of the ethylene-forming enzyme partially purified from melon. Phytochemistry. 1992;31:1485-1494

[25] Harberd NP, Belfield E, Yasumura Y. The angiosperm gibberellin-GID1-DELLA growth regulatory mechanism: How an "inhibitor of an inhibitor" enables flexible response to fluctuating environments. The Plant Cell. 2009;21:1328-1339

[26] Koshioka M. Gibberellin metabolism and its regulation in horticultural plants. Regulation of Plant Growth & Development. 2004;39:1-9 (In Japanese)

Pyridine: A Useful Link for Applications

Chapter 4

Substituent Effect on Pyridine Efficacy as a Chelating Stabilizer

Amer A. G. Al Abdel Hamid

Additional information is available at the end of the chapter

http://dx.doi.org/10.5772/intechopen.75046

Abstract

Owing to the growing interest and unique properties of pyridines as bases, effects of substitution and substituent modification on electron density enrichment of the pyridyl nitrogen, and thus the effectiveness of pyridine as metal ion-stabilizers will be introduced in this chapter. Pyridines of the structure $C_5(S)nH_5\text{-}nN$ (S = substituent) that have been intensively studied theoretically were selected as examples to prove the concept of this chapter. Computational results in the reviewed reports showed that: substitution and substituent modification significantly affect the electronic enrichment of nitrogen atom of the pyridine. The conclusions extracted from the covered investigations were employed to promote pyridines to act as efficient stabilizers for the coordinated metal ions. In coordination chemistry, these kinds of coordinated complexes are highly demanded in the field of chemosensation.

Keywords: pyridine, chelating, substituent effect, DFT-calculation, electron density

1. Introduction

Pyridine is a six-membered N-heterocyclic molecule. It is characterized as relatively strong Brønsted basic [1, 2] and consequently its corresponding pyridinium salt as a relatively weak conjugate acid. Pyridine can serve as a solvent of high donor number [1, 3].

Pyridines compared to carbocyclic analogs have higher nitrogen contents and thus are capable of releasing sufficient delocalized electrons to the system they belong to, and therefore contribute more to the electron-donating activity in the parent compound than C-atom does in carbocyclic compounds. This is attributed to the fact that N-atom possesses a greater number of nonbonding valence electrons compared to C-atom. In effect, the inclusion of pyridine moiety into a molecule is assumed to enrich the molecule with electron density and, correspondingly, improves its stability and binding capability.

In result, and owing to their unique properties as bases, rather high electron density, high positive heat of formation, and good thermal stability [4–7] pyridines and related derivatives are efficiently utilized as chelates (*stabilizing agents*) for the different electron-deficient metal ions in coordination chemistry [1, 8]. Among the very wide spectrum of applications (*e.g. the use in photo-converting systems to energy-rich compounds or to electricity* [9–12]), these synthesized coordination complexes based on pyridines have been extensively employed in photochemical sensation [13–16]. Therefore, they have long been considered the benchmark for understanding many of the photochemical properties of transition metal complexes [1, 17].

Elsewhere in literature [18–22] it has been shown that structural modification of pyridine or pyridine derivatives practices a pronounced effect on photochemistry and dynamics of related complexes. Where, the presence of nitrogen atom in the ligation sphere of the pyridine chelating center, dramatically alters the energy mapping of the HOMO-LUMO (*highest occupied molecular orbital-lowest unoccupied molecular orbital*) and therefore, the energy of the lowest energy intra-ligand charge transfer transitions [13–16, 23].

As such, studying pyridine and/substituted pyridines has attracted the attention of researchers. Where how far, substituent modification (*in terms of number, type, or location of substituent(s)*) on pyridine ring exerts changes on either the close or the more distant neighborhoods of pyridyl-N atom and consequently to what extent this modification has reflected on basicity (*the nitrogen two 2p-type electrons*) of pyridine is chiefly discussed in this chapter.

In one of the related publications [15] coordination complexes of the type: $[Ru(en)_2L_2]_2$, en = ethylenediamine, L = pyridine or substituted pyridine, namely, methyl-, acetyl-, and cyanopyridine, have been theoretically studied. Results indicated that, changing either the type or position of the substituent(s) placed on the coordinated pyridine chelate could facilitate transitions to appear in the visible region. Interestingly, this enables furnishing of colorimetric chemosensors based on the motivated chemo-luminescence properties of the synthesized metal-pyridine inorganic coordination compounds.

Herein, in the following sections, we will introduce some examples that have been discussed in the reported investigations and how the findings in these studies had led to enrich the thorough understanding of how pyridine structure may interrelate to the dynamics of electronic delocalization over pyridine ring and how this, at the end, would recruit modifications in the optical properties of the investigated pyridine derivatives.

In the first study [24] pyridine, C_5H_5N, and pyridine simulations of the type $C_5(X)_nH_{5-n}N$ (X = $-C \equiv C-H$; $-C \equiv C-F$; $-C \equiv N$; $-CH(=O)$) were theoretically studied employing density functional theory (DFT) and time-dependent density functional theory (TDDFT) calculations at the B3LYP/LANL2DZ level of theory.

Gathered results have indicated that substituent type, number, and position had interestingly affected the charge density localization/delocalization on pyridyl nitrogen of the investigated pyridine simulates. For example, substituted pyridine simulates were shown to possess higher stabilization energy compared to that of unsubstituted pyridines. In addition, and regarding the effect of the number of substituents positioned on pyridine ring, more stability was gained when the number of substituents on the ring was increased. For instance, *ortho-*, *meta-*, *para*-trisubstituted C_5H_2N simulate with $-C \equiv C—H$ substituent; abbreviated as $C_5H_2N(o,m,p\text{-}C \equiv CH)_3$ was having greater optimum energy (-476.6 a.u) than the *ortho*, *ortho*-disubstituted $(C_5H_3N(o,o\text{-}C \equiv CH)_2$, optimum energy $= -400.5$ a.u) or *ortho*-monosubstituted $(C_5H_4N(o\text{-}C \equiv CH$, optimum energy $= -324.4$ a.u). This is of course, compared to the unsubstituted pyridine simulate C_5H_5N (optimum energy $= -248.2$ a.u), see **Table 1** and **Figure 1**.

Looking at **Table 1**, one can easily observe the followings:

a. Regardless of the substituent number, the optimum energy of pyridine simulates increases as the number of substituents on the simulate increases. For example, the highest optimum energy is recorded for the pentasubstituted ones **Table 1**.

b. Excluding the substituent number, the effect of substituent type is clearly seen when the substituent $-C \equiv C—H$ was replaced by either $-C \equiv N$ or $-C \equiv C—F$ substituents, where increasing the optimum energy follows the order $-C \equiv C—H < -C \equiv N < -C \equiv CF$. For example, compare $(C_5H_2N(o,m,p\text{-}C \equiv CH)_3$,-476.6 a.u) with $(C_5H_2N(o,m,p\text{-}C \equiv N)_3$,-524.9 a.u) and $(C_5H_2N(o,m,p\text{-}C \equiv CF)_3$,-774.3 a.u).

The observed enhancement in stabilization energy of the pyridine as a result of changing the substituent type is believed to be related to the type of the atom tethered to the tail of the carbon-carbon; $-C \equiv C$ triple bond (*these are: N-atom in $-C \equiv N$ substituent and F-atom in $-C \equiv C—F$ substituent*). In result, this indicates that pyridine is very responsive to even slight changes in the structure of the substituent.

Compound	Optimum energy (a.u)	APT-charge on pyridyl nitrogen (a.u)	Average bond length of *ortho* C=N
C_5H_5N	−248.2	−0.337	1.35828
$C_5H_4N(o\text{-}C \equiv CH)$	−324.4	−0.346	1.36738
$C_5H_3N(o,o\text{-}C \equiv CH)_2$	−400.5	−0.372	1.36326
$C_5H_2N(o,m,p\text{-}C \equiv CH)_3$	−476.6	−0.411	1.36372
$C_5N(o,o,m,p,p\text{-}C \equiv CH)_5$	−628.9	−0.411	1.35747
$C_5H_4N(o\text{-}C \equiv N)$	−340.5	−0.302	1.36253
$C_5H_3N(o,o\text{-}C \equiv N)_2$	−432.7	−0.285	1.35783
$C_5H_2N(o,m,p\text{-}C \equiv N)_3$	−524.9	−0.289	1.35789
$C_5N(o,o,m,p,p\text{-}C \equiv N)_5$	−709.3	−0.346	1.35245
$C_5H_4N(o\text{-}C \equiv CF)$	−423.6	−0.356	1.36598
$C_5H_3N(o,o\text{-}C \equiv CF)_2$	−599.0	−0.396	1.36233
$C_5H_2N(o,m,p\text{-}C \equiv CF)_3$	−774.3	−0.427	1.36276
$C_5N(o,o,m,p,p\text{-}C \equiv CF)_5$	−1125.0	−0.438	1.35648

Table 1. Optimum energy, APT partial charge of pyridyl nitrogen atom and bond length of *ortho* C=N bond in the substituted pyridines compared to that in unsubstituted parent pyridine.

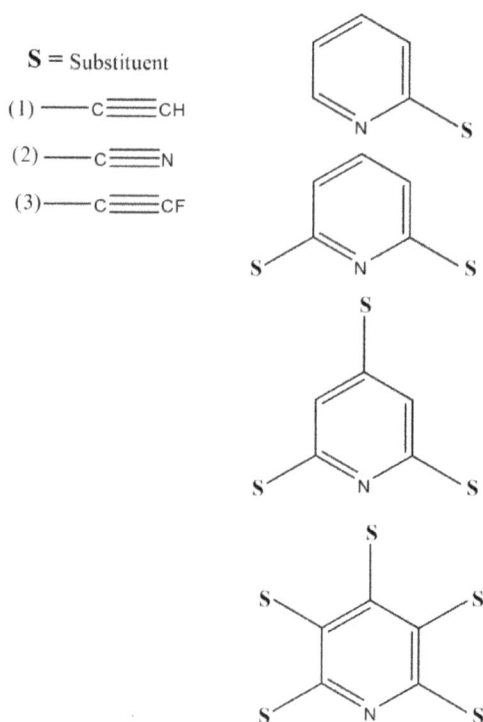

S = Substituent

(1) —— C≡≡CH

(2) —— C≡≡N

(3) —— C≡≡CF

Figure 1. Schematic structures of substituted pyridine simulates, showing the position, number and type of some substituents placed on the ring of the simulated pyridine.

c. Calculating the APT-charge (The Atomic Polar Tensor approach [25, 26]) on pyridyl nitrogen was utilized to monitor the changes in electron density enrichment of pyridyl nitrogen as a result of substituent modification, **Figure 4**. This was very important and beneficial, by being helpful in exploring the effectiveness of pyridyl nitrogen toward binding the targeted electron-deficient metal ion.

Actually calculation results of APT-charge on atoms have shown that various substituents tend to affect the charge density accumulation on pyridyl nitrogen differently. The meant electron density localization/delocalization was inherently found to be highly dependent on the type, number, and position of the substituent(s) attached to the simulate. For instance, –C ≡ C—F compared to –C ≡ C—H and –C ≡ N was found to enrich (or *concentrates*) the charge density on pyridyl nitrogen the most. This, as pointed ahead, was attributed to the existence of the fluorine atom at the tail of the attached –C ≡ C—F substituent. The fluorine atom, through its own p-orbitals, was capable of expanding the electron density delocalization pathway more compared to that of the two substituents; –C ≡ C—H and –C ≡ N, **Table 1** and **Figure 4**.

From the other side, and from coordination chemistry point of view, studying charge density accumulation employing pyridine simulates is essential especially when it is

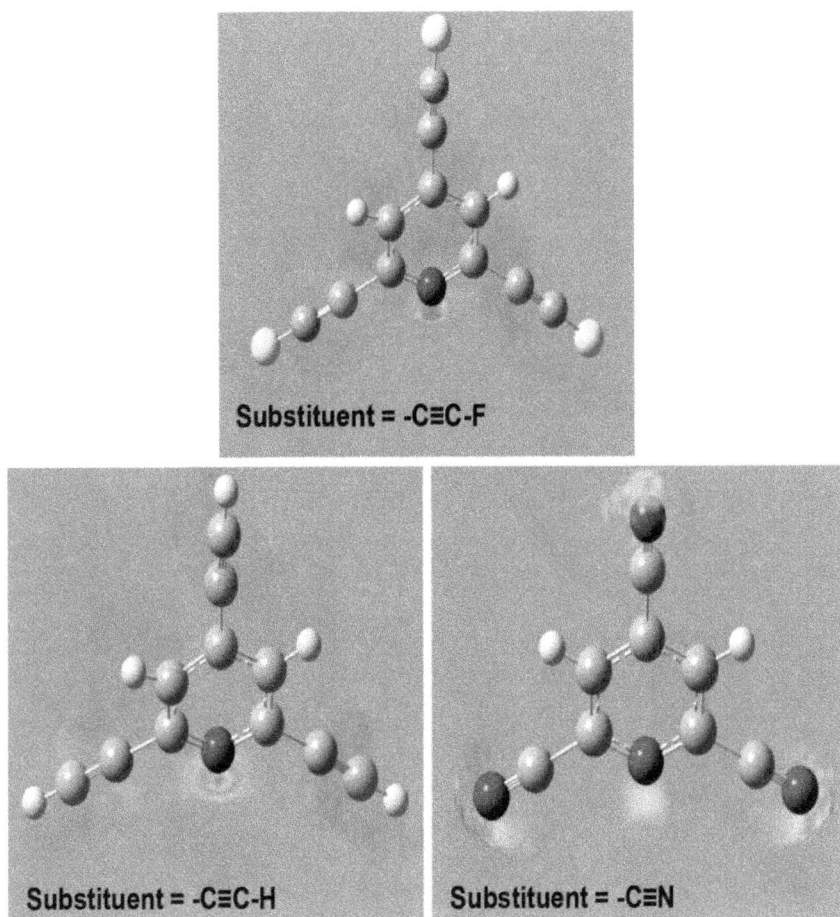

Figure 2. Images of APT-charge distribution of pyridine trisubstituted simulates, showing the variation in charge density on N-atom depending on substituent type. The charge density on N-atom is represented by the intense red-colored spot.

attempted to draw the map of the charge density distribution over atoms constituting the pyridine ring system in general and over pyridyl nitrogen atom in particular. In addition, it also shows how substituent modification interferes with the nitrogen donation capability (*whether reinforced or weakened*) as a result of either dispersing or accumulating the charge density on the nitrogen atom itself or on the other atoms in the close proximity of nitrogen, **Figure 2**.

d. In the referred study, it was reasonable to relate the amount of APT-charge on pyridyl nitrogen to the bond length of the ortho-C=N bond. The key in establishing this relationship between the two is based on the fact that passage of electrons through a given bond affects its force constant. This in effect appears as a change in bond distance of the given bond.

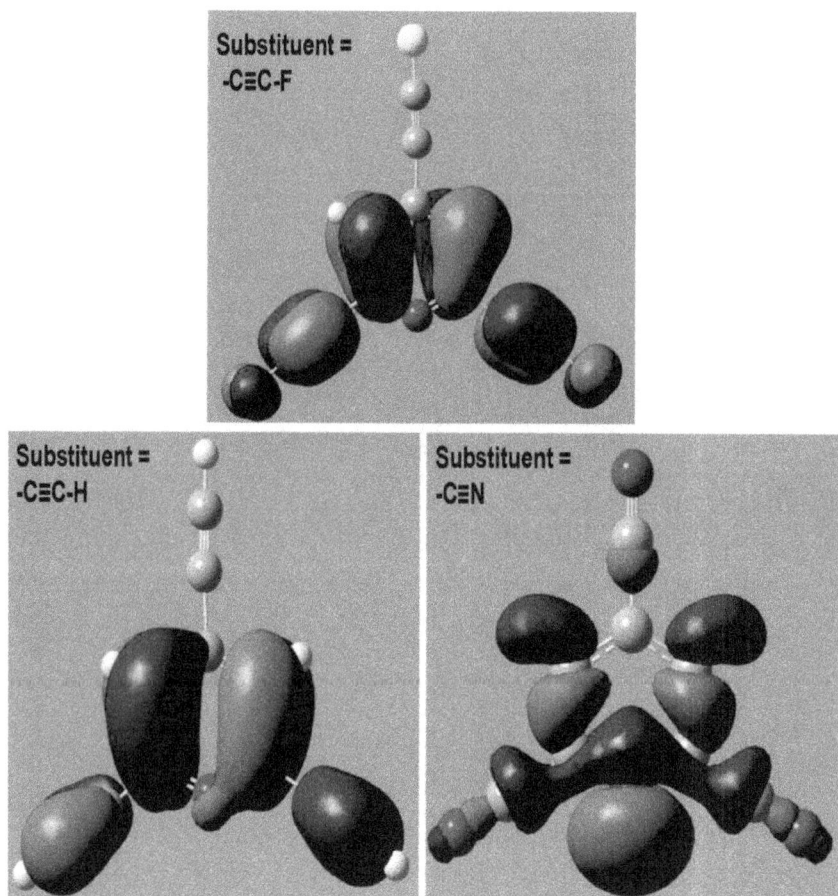

Figure 3. Images of the frontier HOMO molecular orbital component of pyridine trisubstituted simulates, showing the occupation depending on substituent type.

Correlating the values of the *ortho*-C=N bond length with the charge density accumulated on pyridyl nitrogen (see **Table 1**) supports the fact that substituent modification does affect the bond length of the *ortho*-C=N bond in consequence to changing the amount of charge density on pyridyl nitrogen.

Thus, bond length calculations were actually employed to track the variation in charge density in the vicinity of pyridyl nitrogen as a function of substituent modification (**Figure 3**). Why this is important, since it provides deep insights of how charge density is distributed around the pyridyl nitrogen and therefore, how the substituent modification drives nitrogen to be stronger or weaker donor.

Going back to the data presented in **Table 1**, we find that, trisubstituted simulate $C_5H_2N(o,m,p\text{-}C \equiv CH)_3$ of charge density − 0.411 a.u and 1.36372 A^0 bond length shows smaller charge density value and larger bond length relative to the trisubstituted simulate $C_5H_2N(o,m,p\text{-}C \equiv CF)_3$ which has a charge density of-0.427 and 1.36276 A^0 bond length.

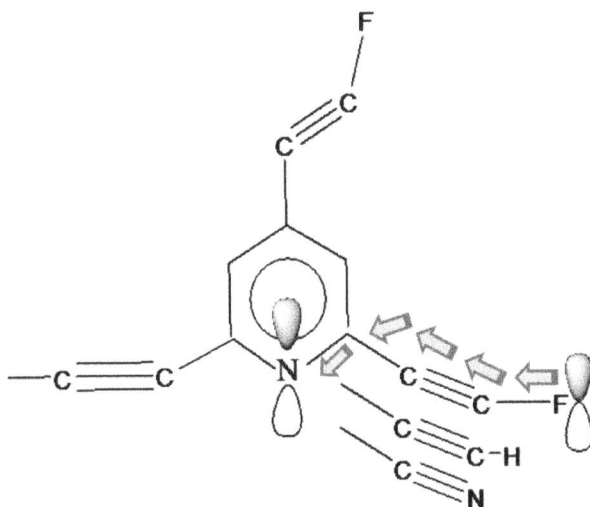

Figure 4. Oversimplified plot showing the assumed extension in the electronic density delocalization pathway generated by the p-orbitals of the fluorine atom on $-C \equiv C-$ substituent compared to H or N-atoms.

From the proceeding, it is obvious that, the most interesting outcomes of the referred study are that, various substituents tend to affect the charge density localization/delocalization differently and found to be highly dependent on the type, number, and position of the substituent attached to the simulate.

Among the investigated substituents ($-C \equiv C-H$, $-C \equiv C-F$, $-C \equiv N$) the substituent $-C \equiv C-F$ was found to be the most efficient in enriching the charge density on pyridyl nitrogen. The fluorine atom [24, 27] existed at the end of the substituent was held responsible for this uniqueness in improving the basicity (*and thus the donating effectiveness*) of the pyridine ring as a whole and pyridyl nitrogen in specific, this is relative to the other two substituents. In effect, the nitrogen atom of the pyridine simulates holding $-C \equiv C-F$ substituent is the one which is promoted the most to act as the efficient stabilizer for the incoming metal ion during coordination process.

Later, and based on the aforementioned findings, one can raise the question: what would be the case if the fluorine atom in $-C \equiv C-F$ substituent is replaced by any other member of the halogen family and how this would affect the basicity of the pyridyl nitrogen. The answer of this questioning came from the outcomes of the study [27] in which calculations on simulates of the type $C_5(S)_n H_{5-n} N$ ($S = -C \equiv C-X$, $X = F$, Cl, Br, I) were conducted.

As it was the case in the studies presented ahead, substituents built up from halogenated carbon–carbon triple bond were introduced onto the pyridine ring in many different forms, for examples some of them were based on monosubstitution at the *ortho*-position and others were introduced onto the pyridyl ring in the form of multisubstituents (*di-*, *tri-* and *penta*-fashions), **Figure 1**.

Responses for substituent modification were monitored by calculating the APT-charge density on nitrogen along with the bond length of the neighboring *ortho* carbon-nitrogen double bond [15, 24, 28–31]. In addition, the distribution of charge density within the simulated system was explored by generating the APT-charge distribution surfaces and the HOMO-molecular orbitals.

Effect of substituent halogenation on electron density enrichment of the pyridine nitrogen, and thus its effectiveness as an electron donor have been investigated. Computational results showed that, substituent halogenation does affect the charge density accumulation on the nitrogen atom of pyridine as well as the $C_2 = N$ bond length, **Table 2**. In addition, the most interesting finding was that charge density localization on pyridyl nitrogen atom has been found to depend on the type of halogen tethered to the triple bond of the attached substituent, this is among other factors. Hardness of the halogen atom attached to the tail of the attached substituent also has been proved to be the determining factor in promoting and qualifying the substituted pyridine to act as an effective electron donor.

The values of the APT-charge on nitrogen were found to increase depending on the substituent order: $-C \equiv C—F < -C \equiv C—Cl < -C \equiv C—Br < -C \equiv C—I$, which is basically the same order of decreasing the hardness of halogens. The maximum enrichment of charge density around nitrogen atom was observed for the iodated triply bonded carbon-carbon substituted pyridines, **Table 2** and **Figure 5**.

Compound	APT-charge on pyridyl nitrogen (a.u)	Bond length of ortho C=N (A°)
C_5H_5N	−0.337	1.35828
$C_5H_4N(o\text{-}C \equiv C—F)$	−0.356	1.36598
$C_5H_4N(o,o\text{-}C \equiv C—F)_2$	−0.396	1.36233
$C_5H_4N(o,m,p\text{-}C \equiv C—F)_3$	−0.438	1.36276
$C_5H_4N(o,o,m,p,p\text{-}C \equiv C—F)_5$	−0.427	1.35648
$C_5H_4N(o\text{-}C \equiv C—Cl)$	−0.365	1.36691
$C_5H_4N(o,o\text{-}C \equiv C—Cl)_2$	−0.419	1.36296
$C_5H_4N(o,m,p\text{-}C \equiv C—Cl)_3$	−0.476	1.36349
$C_5H_4N(o,o,m,p,p\text{-}C \equiv C—Cl)_5$	−0.469	1.35714
$C_5H_4N(o\text{-}C \equiv C—Br)$	−0.372	1.36732
$C_5H_4N(o,o\text{-}C \equiv C—Br)_2$	−0.436	1.36315
$C_5H_4N(o,m,p\text{-}C \equiv C—Br)_3$	−0.502	1.36384
$C_5H_4N(o,o,m,p,p\text{-}C \equiv C—Br)_5$	−0.493	1.35751
$C_5H_4N(o\text{-}C \equiv C—I)$	−0.380	1.36752
$C_5H_4N(o,o\text{-}C \equiv C—I)_2$	−0.454	1.36351
$C_5H_4N(o,m,p\text{-}C \equiv C—I)_3$	−0.530	1.36411
$C_5H_4N(o,o,m,p,p\text{-}C \equiv C—I)_5$	−0.519	1.35778

Table 2. APT partial charge of pyridyl nitrogen atom and bond length of *ortho* C=N bond in the halogenated triply-C≡C-bonded substituted pyridines compared to that in unsubstituted parent pyridine.

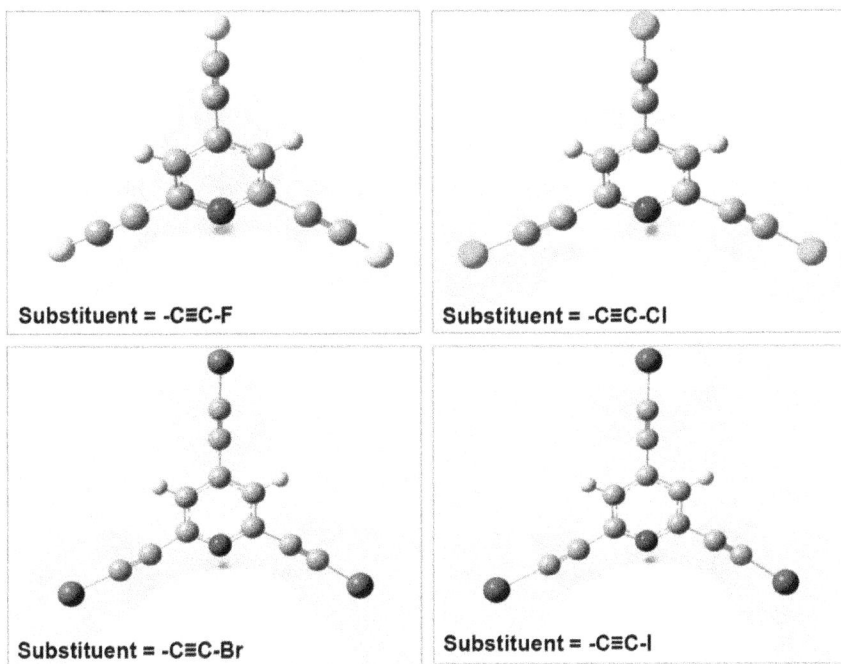

Figure 5. Images of APT-charge distribution of halogenated triply -C ≡ C- bonded trisubstituted pyridines, showing the variation in charge density on N-atom depending on substituent type.

The conclusions extracted from the referred [27] study support the proceeded findings of the earlier studies [15, 24], where deeper insights on the key factors that may qualify chelates to function as good stabilizers for metal ions in coordination complexes were gained. This of course is very valuable since it is highly demanded in the field of chemosensation.

As seen in **Table 2**, and in contrast to other analogs, the iodated trisubstituted simulates show the highest accumulation of electron density around nitrogen, indicating that much higher charge density was accumulated on nitrogen by iodine (*the softest halogen*) being tethered at the tail of the carbon-carbon triple bond of the substituent on pyridine. This is compared to the other members of halogens.

In the section of final conclusions drawn in the aforementioned study, the hardness of halogen atom attached to the tail of the substituent undoubtedly was held responsible for controlling the amount of charge density localized on the nitrogen atom of the simulate. This amount of charge density on nitrogen increases as halogen atom changes in the order F, Cl, Br and I, **Table 2**.

In this investigation, it was quite interesting to witness the halogen atom (*which is tethered to the triple bond at the far end of the substituent on pyridine*) masters the donation effectiveness of the pyridyl nitrogen and thus may promote or prohibit the tendency of the pyridine chelate to act as an efficient stabilizer for the targeted electron-deficient metal ion. Wise investing of this property in

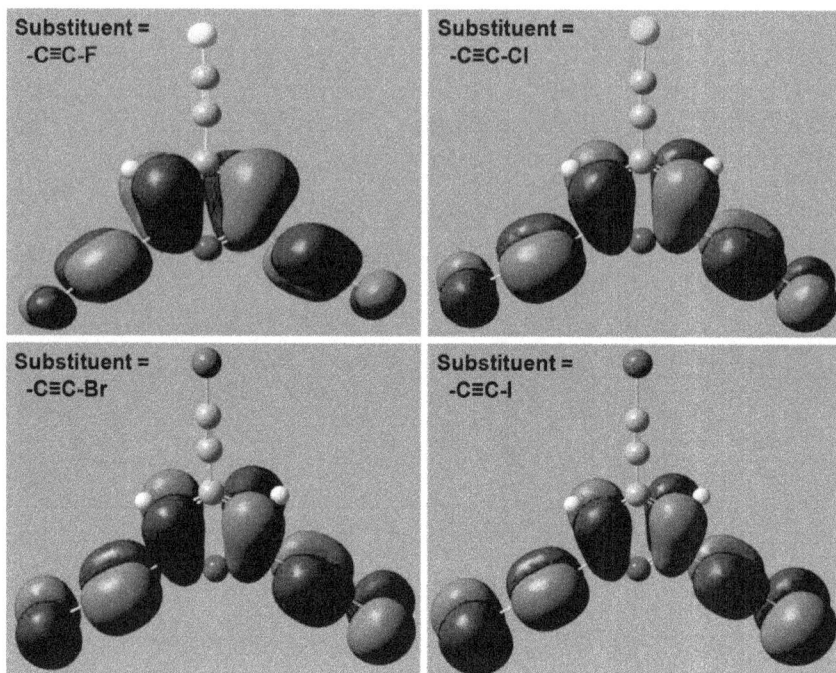

Figure 6. Images of the frontier HOMO molecular orbital component of halogenated triply -C ≡ C- bonded trisubstituted pyridines, showing the occupation depending on the substituent type.

analogous systems would enable synthesizing new generations of pyridine derivatives that are hopefully suitable for engineering a low-energy absorbers, the type of inorganic coordination complexes that are essential and highly demanded in the field of colorimetric chemosensation.

Changes in bond distance of the $C_2 = N$ bond were relied on as extra supporter for the suggested migration image of charge density toward halogen and away from nitrogen, **Table 2**.

This was acceptable, since any variation in the charge density localization on nitrogen would be accompanied by a direct reflection on the $C_2 = N$ bond length. This is equivalent to say that, the more the charge density is localized on nitrogen atom, the less the charge density is swept away from nitrogen or passed through the $C_2 = N$ bond, meaning that weaker (*smaller force constant*) bond will be. For example, trisubstituted simulates (*which have recorded the highest values of charge density enrichment on nitrogen*) are found to exhibit $C_2 = N$ bond lengthening compared to free pyridine, **Table 2**. In other words, this means that, introduction of the halogenated substituent(s) in the form of trisubstitution described in **Figure 1** has led the charge density to accumulate on nitrogen and therefore effectively improved the donation capability of pyridyl nitrogen.

Looking at the shapes of HOMOs (*highest occupied molecular orbitals*) shown in **Figure 6** enables mapping the charge density distribution on atoms of the pyridine moiety.

As seen in the figure, the HOMO of the simulate bearing the -C ≡ C—F substituent extends over the C_2-C_3, C_5-C_6 atoms including nitrogen and the loops are notably larger in size compared to that observed for the other simulates which are based on Cl-, Br- and I-halogenated substituents. The HOMO loops were reduced in size as the halogen atom changes in the order F, Cl, Br and I, indicating the dependence of the HOMO expansion on the hardness of the halogen atom attached to the substituent. Noticeably, as more charge density is withdrawn from pyridyl nitrogen toward halogen, larger-sized HOMO loops are anticipated, for example compare $-C \equiv C-F$ simulate with $-C \equiv C-I$ simulate, **Figure 6**.

The same argument is applied, when the APT-charge density is correlated to the absorption energy of the simulate. In the same study, it has been pointed that, as more charge density is localized on nitrogen, a pronounced red shift in the absorption energy of the simulate is observed. This reveals that HOMO/LUMO molecular orbitals practice a narrow energy gap, and therefore simulates of higher APT-charge density on nitrogen are expected to show longer wavelength energy bands of absorption. This finding is considered as the most interesting and advantageous finding in the referred study, where it suggests the possibility of controlling the absorption energy of a particular simulate by playing with the hardness of the halogen atom attached to the tail of the attached substituent(s) on pyridines.

2. Conclusions

The overall conclusion that can be deduced from the aforementioned review is that various substituents can exert pronounced and beneficial effects on charge density enrichment of the pyridyl nitrogen atom in pyridines.

This in result allows stating that substituent modification (*in the form of number, type or position of substitution*) can be employed as a tool for controlling the donation effectiveness of the nitrogen atom in pyridines.

These achievements are essential, when inorganic complexes are demanded to be utilized in the field of chemosensation to feasibly engineer low-energy optically active luminescent inorganic compounds.

Conflict of interest

The author declares that he has no conflict of interest regarding publication of this chapter.

Author details

Amer A. G. Al Abdel Hamid

Address all correspondence to: amerj@yu.edu.jo

Department of Chemistry, Yarmouk University, Irbid, Jordan

References

[1] Krygowski TM, Szatyłowicz H, Zachara JE. How H-bonding Modifies Molecular Structure and ð-Electron Delocalization in the Ring of Pyridine/Pyridinium Derivatives Involved in H-Bond Complexation. The Journal of Organic Chemistry. 2005;**70**:8859

[2] Kenneth RA. New Density Functional and Atoms in Molecules Method of Computing Relative pKa Values in Solution. The Journal of Physical Chemistry. A. 2002:**106**(49): 11963

[3] Marcus YJ. The Effectivity of Solvents as Electron Pair Donors. Solution Chemistry. 1984; **13**:599

[4] Huynh M, Hiskey M, Ernest L. Polyazido High-Nitrogen Compounds: Hydrazo- and Azo-1,3,5-triazine. Angewandte Chemie, International Edition. 2004;**43**:4924

[5] Li XT, Pang SP. Luo YJ. New Advances in N-Oxidation of Nitrogen-Containing Heterocyclic Compounds. Chinese Journal of Organic Chemistry. 2007;**27**:1050

[6] Zhou Y, Long XP, Wang X. DFT Studies on the Tetrazine Substituted by Six-membered C−N Heterocyclic Derivatives. Chinese Journal of Energetic Materials. 2006;**14**:315

[7] Liu H, Wang F, Wang G-X, Gong X-D. Theoretical studies of -NH$_2$ and -NO$_2$ substituted dipyridines. Molecular Modeling. 2012;**18**:4639

[8] Institute for Scientific Information. Philadelphia (1996-2005); Retrieved June 2005

[9] Johansson E et al. Spin-Orbit Coupling and Metal-Ligand Interactions in Fe(II), Ru(II), and Os(II) Complexes. Journal of Physical Chemistry C. 2010;**114**:10314

[10] Gratzel M. Photoelectrochemical Cells. Nature. 2001;**414**:338 DOI: 10.1038/35104607

[11] Hagfeldt A, Gratzel. M. Molecular Photovoltaics. Accounts of Chemical Research. 2000; **33**:269

[12] Acevedo JH, Brennaman MK, Meyer T. Chemical Approaches to Artificial Photosynthesis. Journal of Inorganic Chemistry. 2005;**44**:6802

[13] Amer A. Hamid et al. A Selective Optical Chemosensor for Iron(III) Ions in Aqueous Medium Based on [bis(2,2`-bipyridine)-bis(2-carboxypyridine)] ruthenium(II) Complex. Jordan Journal of Chemistry. 2011;**6**(4):393

[14] Amer A. Hamid et al. International Journal of Inorganic Chemistry 2011;**2011**:6. Article ID 843051. DOI: 10.1155/2011/843051

[15] Hamid AA. Research on Chemical Intermediates. 2012;**38**(9)

[16] Hamid AA, Kanan S. Journal of Coordination Chemistry. 2012;**65**(3):420

[17] Seddon KR. Coordination compounds. Coordination Chemistry Reviews. 1982;**41**:79

[18] Juris A et al. Coordination Chemistry Reviews. 1988;**84**:85

[19] Roundhill DM. Photochemistry and Photophysics of Metal Complexes: Modern Inorganic Chemistry Series. New York: Plenum Press; 1994

[20] Kalyanasundaram K. Photochemistry of Polypyridine and Porphyrin Complexes. New York: Academic Press; 1992

[21] Dovletoglou A, Adeyemi SA, Meyer TJ. Inorganic Chemistry. 1996;35:4120

[22] Liu Y et al. Ru(II) Complexes of New Tridentate Ligands: Unexpected High Yield of Sensitized O$_2$. Inorganic Chemistry. 2009;48:375

[23] Bryan Sears R, Joyce LE, Turro C. Electronic tuning of Ruthenium complexes with 8-quinolate ligands Photochemistry and Photobiology. 2010;86:1230

[24] Hamid A, Kanan S, Tahat ZA. DFT analysis of substituent effects on electron-donating efficacy of pyridine. Research on Chemical Intermediates. 2014. DOI: 10.1007/s11164-014-1783-6

[25] Cioslowski J. A new population analysis based on atomic polar tensors. Journal of the American Chemical Society. 1989;111:8333

[26] Gross KC, Seybold PG, Hadad CM. Comparison of different atomic charge schemes for predicting pKa variations in substituted anilines and phenols. International Journal of Quantum Chemistry. 2002;90:445

[27] Abdel Hamid A et al. Chemosensor Engineering: Effects of Halogen Attached to Carbon-Carbon Triple Bond Substituent on Absorption energy of Pyridine: DFT-Study. Jordan Journal of Chemistry. 2016;11(1):8

[28] Manz TA, Sholl DS. Improved Atoms-in-Molecule Charge Partitioning Functional for Simultaneously Reproducing the Electrostatic Potential and Chemical States in Periodic and Nonperiodic Materials. Journal of Chemical Theory and Computation. 2012;8(8):2844

[29] Dunning TH, Hay PJ. Modern Theoritical Chemistry. Schaefer. In: H. F III Ed, editor. New York: Plenum; 1976. 1-28

[30] Cramer CJ. Essentials of Computational Chemistry: Theories and Methods. Wiley; 2002. pp. 278-289

[31] Heinz H, Suter UW. Atomic Charges for Classical Simulations of Polar Systems. The Journal of Physical Chemistry. B. 2004;108:18341

Chapter 5

Pyridine: A Useful Ligand in Transition Metal Complexes

Satyanarayan Pal

Additional information is available at the end of the chapter

http://dx.doi.org/10.5772/intechopen.76986

Abstract

Pyridine (C_5H_5N) is being the simplest six-membered heterocycles, closely resembles its structure to benzene. The "N" in benzene ring has its high electronegativity influence on resonance environment and produces markedly different chemistry from its carbon analog. The presence of nitrogen and its lone pair in an aromatic environment makes pyridine a unique substance in chemistry. The sp^2 lone pair orbital of "N," directed outward the ring skeleton, is well directed to have overlap with vacant metal orbital in producing an σ bonding interaction. This causes pyridine to be a ligand and has been utilized with all transition metals in producing the array of metal complexes. A rich literature of metal complexes is now available with pyridine and its derivatives. Innumerable complexes have been synthesized with academic as well as industrial importance. To shed a light on ligating capability of pyridine, transition metal complexes with pyridine, and its derivative is presented in this chapter.

Keywords: pyridine, ligand, synthesis, metal complex

1. Introduction

Pyridine is one of the simplest heterocycle known since its discovery in 1849 by Scottish chemist Thomas Anderson. It closely resembles with benzene structure, where a benzene methine (=CH-) group is occupied by "N" to form a six membered aromatic heterocycle of formula C_5H_5N. It is a room temperature colorless, water-soluble liquid with the distinctive pungent smell. The presence of electronegative "N" in ring structure is the sole cause of new properties induced in pyridine differentiating it markedly from benzene. The "N" presence in ring prevents the electron density be distributed evenly over the ring owing to its negative inductive effect, which also causes the weaker resonance stabilization (117 kJmol⁻¹) than benzene (150 kJmol⁻¹) [1]. This is evident from the shorter C-N bond distance (137 pm) compare to 139 pm of the C-C bond. The other bond lengths satisfy typical aromatic nature of the pyridine ring. Similar to benzene all the pyridine ring atoms are sp^2 hybridized and involve in the π electron resonance. The sp^2 "N" involvements in resonance come through its unhybridized p-orbital rather than involving its lone pair. The lone pair thus in sp^2 orbital remain directed outward of the ring in the same plane without contributing to the aromatic behavior of pyridine but greatly influence the chemical environment of the ring. The available "free" lone pair thus could be utilized by "N" in several ways suiting for chemical reactions either on pyridine ring or as Lewis base to form coordinate bond with Lewis acids (**Figure 1**). It is a weak base reacts with acids to get protonated to pyridinium salt with pK_a of conjugate acid (pyridinium cation) is 5.25. For an illustrative example the pyridine reacts with p-toluenesulfonic acid and gets protonated to pyridinium p-sulfonate salt. The protonated pyridine thus produced is isostructural and isoelectronic with benzene.

The pyridine owing to its Lewis basic character rooted in its nitrogen lone pair qualifies as the ligand for transition metals and able to form metal complexes across the metals in the periodic table. It is usually a weak monodentate ligand having capability to bind metal in different proportions to produce the range of metal complexes. A rich literature of pyridine coordinated complexes of transition metals has grown over the years. Pyridine and its numerous derivatives have been under investigation of inorganic chemists in design and preparation of numerous metal complexes of their interest. The further design of pyridine ligand explored in polypyridine system fusing two or more pyridine moieties to result in chelating multidentate ligands. Such as bipyridine, a fused two pyridine rings system, is a worth mention in transition metal chemistry. The unique photochemistry and luminescent ruthenium bipyridine complexes are a glimpse of bipyridine metal chemistry. Phenanthroline, terpyridine, and other multidentate ligands have been in focus in the transition metal chemistry since long. The discussion of bipyridine and polypyridine complexes is beyond the scope of this discussion. In this chapter, we will have a look of ligating capability of basic pyridine unit to transition metal ions.

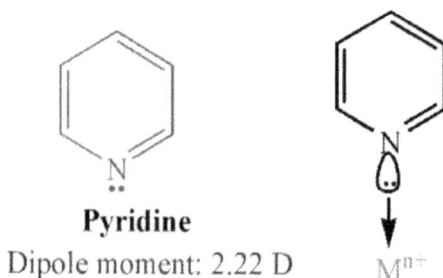

Pyridine
Dipole moment: 2.22 D

Figure 1. Structure of pyridine and its ligation to metal.

2. Pyridine and its metal complexes

2.1. Ligating power of pyridine

The spectrochemical series of ligands in crystal field theory (CFT), portray ligand arrangement pertaining to their metal d-orbital splitting capability, depicts pyridine as moderately strong ligand [2]. This interprets to strong electrostatic interactions of pyridine lone pair to metal d- orbitals. Despite being neutral, pyridine causes moderately large d-orbital splitting implying to strong bonding interaction to metal centers. Beside the CFT, the valence bond theory (VBT), considers metal pyridine bonding to overlap of sp^2 lone pair orbital of pyridine to hybridized metal orbitals. The extent of overlaps is happened to be highest in first transition metals in comparison to the second and third transition elements owing to the difference of shape, size, and energy of the combining orbitals. Apart from nitrogen lone pair orbitals, the ring π-electron is also capable of bonding interaction to metal ions. Moreover delocalized π^* anti-bonding orbitals can act as acceptor of metal electron density (**Figure 2**). The pyridine can also indulge in hydrogen bonding and π-π stacking-like weak interactions. Thus pyridine is enriched with multiple orbitals for bonding interactions with metal ions.

2.2. Pyridine transition metal complexes

The pyridine transition metal complexes have a rich literature. Pyridine found to coordinate all the transition metals producing the variety of metal complexes in their different oxidation states. Efforts were made to incorporate the increasing number of pyridines in metal coordination sphere but exclusive pyridine complexes such as $[M(py)_4]^{n+}$ or $[M(py)_6]^{n+}$ are rare (where M = transition metal). The metal- pyridine chemistry incorporates pyridine and its derivatives with the capability of bidentate or tridentates ligands in the formation of metal complexes. The discussion here to be restricted to the domain of pyridine and it coordination to transition metals. A brief overview of pyridine transition metals complexes is presented here.

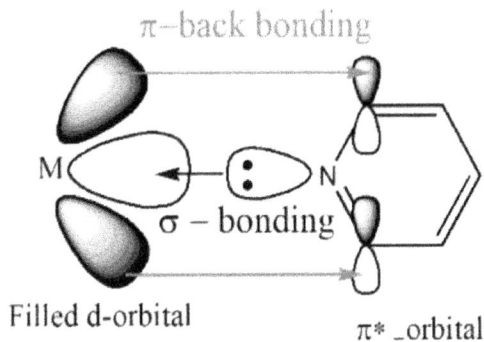

Figure 2. Bonding picture of pyridine.

Scandium and yttrium preferably bind to three pyridine units in a four-coordinated geometry. The coordination number might vary with characteristics of binding ligands. Five and six coordinated complexes are also synthesized in combination of pyridine and thiocyanate (SCN) ligands. The pyridine derivative capable of acting as a bidentate ligand, such as picolinic acid, prefers to produce higher coordination number complexes. A good number of complexes are known with variously substituted pyridines. These complexes are known in +1 and + 3 states of Sc and Y. The example of pyridine coordinated complexes include [Sc(py)$_3$Cl] [3, 4], [Y(py)$_3$Cl] [5, 6], [Sc(py)$_2$(NCS)$_3$] [7], and [Sc(py)$_3$(NCS)$_3$] [8]. These complexes could be derived directly from their metal salt and pyridine at room temperature.

Pyridine complexes of titanium are usually found in Ti(IV) state but could exist in other valence states (II-IV) also. Zirconium also produces pyridine complexes in Zr(III) and Zr(IV) states. Pyridine complexes of Titanium and Zirconium in +4 states are more common. Ti(II) and Ti(IV) pyridine complexes could be prepared along with halogen ligands with formula [Ti(py)$_2$(Cl)$_4$], [Ti(py)$_2$(Br)$_4$] [9]. The Ti(II) complexes use to be tetra-coordinated and those of Ti(III) and Ti(1 V) are hexa- and octa-coordinated [10]. Zr(IV) pyridine complexes have the similar behavior of titanium. A commonly known complex is [Zr(py)$_2$Cl$_4$] [11]. Three representative complexes are depicted in **Figure 3**.

The Ti(II), Ti(III), and Zr(III) pyridine complexes required preparation under inert and moisture free environment. The presence of aerial oxygen facilitates decomposition and formation of +4 states of metal centers. This is evident from preparative methods of Ti(IV) and Zr(IV) pyridine complexes under oxygen. In general, simple pyridine complexes are sensitive to moisture and air [12]. The electron donating substituents at two and four positions help "N" in forming the stronger coordinate bond and enhance the stability of resultant complexes. Beside the chelate pyridine ligands provide appreciably higher stability compare to the monodentate pyridine moiety.

The vanadium, niobium, and tantalum possess rich chemistry of pyridine complexes. The richness arises due to numerous valence states (0-V) of vanadium, which are found to exist with different pyridine complexes. Though pyridine complexes with V (0) and V (I) are very rare. Example of V (0) is [V$_2$ (2-Me-py)$_4$(CO)$_{12}$] [13]. A handful of complexes with other oxidation states are reported. V (II) pyridine complexes were prepared with a range of monodentate and bidentate anionic ligands. Along with basic pyridine moiety, various substituted derivatives were found in the coordination sphere of V (II) centers [14–16]. The simplest

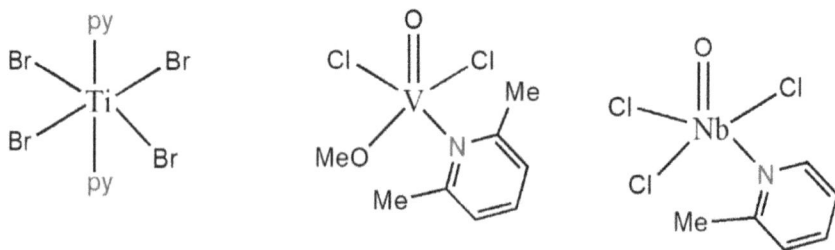

Figure 3. Representative pyridine complexes of Ti, V and Nb.

example of V (II)-pyridine combination includes [V^{II}(py)$_2$(acac)$_2$]. A similar combination of anionic monodentate and chelate ligands could produce V (III) pyridine complexes, such as [V(py)(NCS)$_3$] [17]. The inclusion of oxo group helped in stabilizing V(IV) and V(V) states and a range of pyridine complexes were synthesized. A few illustrative examples are [V^{IV}(O) (4-Et-py)$_2$(acac)$_2$] [14], [V^{V}O(2,6-Me$_2$-py)(OMe)(Cl)$_2$] [18] (**Figure 3**) and so on. Niobium and Tantalum produce pyridine complexes at +4 and + 5 oxidation states. [Nb(4-Me-Py)$_2$(Cl)$_4$] [19], [Nb(py)(OMe)$_5$] [20], [Nb(O)(2-Me-py)(Cl)$_3$] [21], and so on, are few niobium pyridine complexes. Similar combination of Ta complexes are known with halide ligands, for example, [Ta(py)$_2$X$_4$] [22] (X = halide), [Ta(py)(OMe)$_5$] [23] etc.

The complexes could be synthesized from halides salts of respective metals and pyridine in a neutral environment. The tetravalent and pentavalent complexes are harvested in a low-temperature environment to prevent decomposition through disproportionation. The higher valent pyridine complexes possess superior stability than their lower valent counterpart.

Chromium, molybdenum, and tungsten pyridine complexes could be obtained from their inorganic salts as well as carbonyl and nitrosyl complexes. Neutral environment remain a preferred choice to ensure the stability of synthesized complexes. The large range of oxidation states (0-VI) of these metals has produced innumerable pyridine complexes. The tendency of lower valent pyridine complexes, particularly in 0 and + 1 states, get oxidized to higher oxidation states made these complexes sensitive to air and moisture. The common coordination number remains six in Cr and Mo, but it could be in higher numbers in tungsten pyridine complexes.

Cr(0) and Cr(I) pyridine complexes are accompanied with carbonyl and nitrosyl ligands, such as [Cr(py)(C$_5$H$_5$)(NO)(CO)] [24] (**Figure 4**). The higher valent pyridine complexes have anionic monodentate and bidentate ligands in Cr(III) and Cr(IV) complexes. The Cr(IV) and upper oxidation state mostly found with an oxo group. The representative chromium-pyridine complexes are [Cr(py)(acac)$_3$] [25], [Cr(O)(py)(Br)$_3$] [26], [Cr(O)$_3$(py)] [27] and so on, reflects the above facts. The oxo group continues to stabilize higher valent molybdenum and tungsten pyridine complexes too. This is evident from the formula of [Mo(py)(O)$_3$], [Mo(NCS)$_2$(O)$_2$] [28], [W(py)(O)$_2$(Cl)] [29], [WO(py)(Cl)$_4$] [30] (**Figure 4**) and so on. The +6 state of tungsten-pyridine complexes could also be stabilized by monodentate anionic ligands as it found in [W (py)F$_6$] [31], and [WO(py)Cl$_4$] [30].

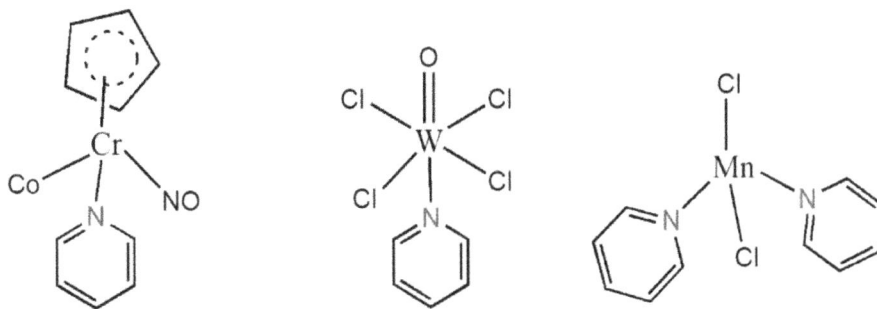

Figure 4. Representative pyridine complexes of Cr, W and Mn.

Manganese and rhenium forms complexes with pyridine in different oxidation states spreading over 0 to VII. The coordination number commonly varied from four to eight. However, the manganese forms pyridine complexes only in zero to quadrivalent oxidation states, whereas rhenium pyridine complexes exist in seven oxidation states. The lower valent pyridine complexes of these metals are composed of carbonyl and nitrosyl counterpart. The higher valent rhenium accommodates oxo ligands along with anionic monodentate and chelating ligands. The Mn(I) complexes quickly react with air and oxygen. Thus their preparation is carried out in a neutral atmosphere. The higher valent pyridine complexes are stable in normal condition and could be prepared in alcoholic or aqueous media. Manganese halides and manganese oxide are remaining preferred starting materials for synthesizing pyridine complexes. Complexes such as $[Mn(py)_2X_2]$, (X = halide) (**Figure 4**), $H[Mn(H)(py)(Cl)_3(OH)]$, and $[Mn(py)_2(thiourea)_4Cl_2]$ [32–34] were obtained from these starting materials. The rhenium-pyridine complex preparation has also origin at rhenium halides, such as ReI_4, $K_2[ReCl_6]$ are few to mention. The higher valent rhenium pyridine compound also derived from $K[ReO_4]$. The examples of rhenium complexes [35–38] are $[Re(4-Me-py)(Br)(CO)_4]$, $[Re(py)Cl_2(CO)_2(NO)]$, $[Re(py)(O)(Cl)_4]$ (**Figure 5**), $[Re(py)_2(OH)_2(Cl)_3]$ and so on. Technetium-pyridine complexes are rare [39, 40].

Fe(II) has produced quite a significant number of pyridine complexes with the comparison to other first transition metals. A range of pyridine derivatives were included in these Fe-pyridine complexe. The number of pyridine complexes of Fe(II) are quite high in compare to Fe(0), Fe(III), and Fe(IV). This is due to the higher stability of Fe(II) pyridine complexes than the relevant complexes of Fe(0), Fe(III), and Fe(IV). In case of iron, pyridine coordination to metals center achieved both in mixed ligand and fully pyridine coordinated environment, such as $[Fe(py)_6]^{2+}$ [41] (**Figure 5**). Fe(II), Fe(III), and Fe(IV) pyridine complexes are mostly octahedral. Though four and five coordinated complexes are also seldom found, the pyridine complexes were prepared by interaction of pyridine and an inorganic salt of iron. Few representative iron pyridine complexes [42, 43] of different oxidation states are $[Fe(CO)_4(py)]$, $Na_3[Fe(2-NH_2-py)(CN)_5]$, $[Fe(py)_4I_2]$ and so on.

In this group, the versatility of complex formation continues with ruthenium and osmium too. This is evident from numerous pyridine complexes reported with these metals. Ruthenium displays nine oxidation states (0-VIII). Among these +2 and +3 oxidations are most stable. The pyridine complexes of ruthenium contain +I, +II, +III, +IV, +VI, and + VIII states. Few complexes even

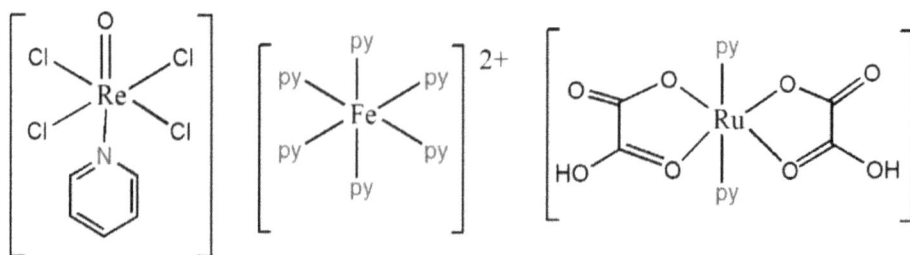

Figure 5. Representative pyridine complexes of Re, Fe and Ru.

reported with fractional oxidation number. Like ruthenium, osmium pyridine complexes found with II-IV, VI, and VIII oxidation states. The common coordination number in such complexes ranges from 4 to 6, though higher coordination numbers are claimed in higher oxidation states.

Ruthenium pyridine complex preparation involves high-temperature reflux of ruthenium salt with pyridine in an organic solvent preferably in the oxygen-free environment. $RuCl_3$, $[Ru(NH_3)_6]Cl_2$, "ruthenium red" $[(NH_3)_5Ru-O-Ru(NH_3)_4-O-Ru(NH_3)_5]^{6+}$ are the common choice for the synthesis of pyridine complexes. These resultant pyridine complexes are often labile and subject to the decomposition. Ru(III) pyridine complexes often show up dispro-portion to Ru(II) and Ru(IV) states and this decomposition route appears to be a synthetic procedure for new complex preparation. Few ruthenium complexes [44–47] with varied pyridine numbers are $K[Ru(py)(Cl)_4]$, $K[Ru(py)_2(ox)_2]$ (**Figure 5**), $[Ru(O)_4(py)_2]$, $[Ru(py)_6]$ $(BF_4)_2$ and so on. The coordination of six pyridine ligands to ruthenium, $[Ru(py)_6](BF_4)_2$, is a unique example of "pure" pyridine complex, which is infrequent in metal-pyridine chemistry.

The osmium pyridine complex preparations could be achieved by reaction of $K_2[OsCl_6]$ and pyridine. OsO_4 also proved to react with pyridine and produce higher valent pyridine com-plexes. In a reaction of $[Os_3(CO)_{12}]$ with pyridine in neat or with a pyridine saturated hydro-carbon solvent resulted series of complexes [48], such as $[HOs_3(py)(CO)_{10}]$, $[HOs_3(py)(CO)_9]$, $[H_2Os_3(py)_2(CO)_8]$, $[Os_2(py)_2(CO)_6]$, and so on, $K[OsCl_4(bpy)]$ gave rise to pyridine complex $[Os(py)Cl_3(bpy)]$ by treatment with aqueous pyridine in boiling condition or treatment in pyri-dine–glycerol mixture [49]. In Glycerol pyridine mixture $[OsCl_4(bpy)]$ complex reduced to pro-duce $[Os(py)Cl_3(bpy)]$ (**Figure 6**) and $[Os(py)_3Cl_2(bpy)]ClO_4$. These complexes are fairly stable except the labile Os(II) complexes.

Cobalt pyridine complexes were isolated with Co(II) and Co(III) oxidation states. Only few Co(I) and Co(IV) complexes are reported. The Co(II) could accept one to six pyridine ligands in its coordination sphere and the resultant complexes are mostly six coordinated. One such complex contains six pyridine ligands (**Figure 6**) and formulated as $[Co (py)_6]I_2$ [50]. Most of the cobalt pyridine complexes are mixed-ligand molecule.

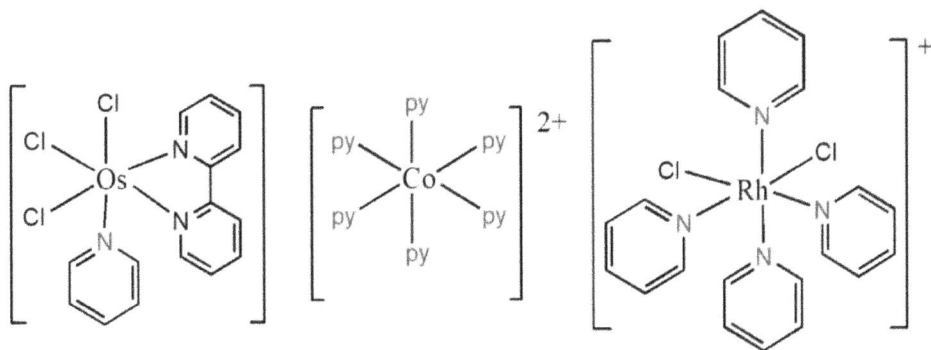

Figure 6. Representative pyridine complexes of Os, Co and Rh.

They could either prepared from alcoholic or aqueous solution with reaction of cobalt salt and appropriate quantity of pyridine. When cobalt iodide treated with excess of 3-ethyl pyridine a brown complex [Co(3-Et-py)$_4$Cl$_2$] resulted [51]. Several other substituted pyridine employed in a similar fashion resulting in numerous complexes. Cobalt(I) complexes usually resulted from reduction of Co(II)-pyridine complexes.

Rhodium also used its various oxidation states to have pyridine coordination. Pyridine complexes are known with +1 to +4 and +6 oxidation states with coordination numbers four to six. Rh(III) dominates the spectra of pyridine complexes with octahedral geometrical preference, whereas square planar geometry is common finding with Rh(I) complexes. RhCl$_3$ is the most common starting material for preparation of pyridine complexes. [RhL$_4$X$_2$]$^+$ (where L = 3 or 4 substituted pyridine, X = halide) type complexes were readily synthesized from rhodium halide and pyridine interaction in aqueous medium [52]. [Rh (PPh$_3$)$_2$(CO)(Cl)] is another starting material for preparation of Rh(I) pyridine complexes.

Iridium also displays eight different oxidation states (−1 to VI) and pyridine complexes are favorably formed with +3 state. Ir(I) complexes can have four and five coordination, whereas Ir(III) and Ir(IV) can extend their coordination number to six. The preparation of iridium pyridine complexes can be achieved from the array of starting materials. Ir$_2$(SO$_4$)$_3$, K$_3$IrX$_6$ (X = halide), K$_2$[Ir(H$_2$O)Cl$_5$] are few to mention. The number of pyridine moiety varies around iridium center and it could reach maximum six. Iridium(IV) pyridine complexes derived from Ir(III) counterpart by oxidation. Few illustrative examples of iridium pyridine complexes [53, 54] are K [Ir(py)$_2$Cl$_4$], [Ir(py)$_3$Cl$_3$] (**Figure 7**), [Ir(py)$_3$(H$_2$O)Cl$_2$] and so on.

The pyridine nickel complexes are mostly found in +2 oxidation states. The other oxidation states, that is, Ni(I), Ni(III), and Ni(IV) are less numerous in nickel-pyridine chemistry. Ni(II) complexes could be tetra, penta, and hexacoordinated. One such octahedral complex is [Ni(py)$_4$X$_2$] [55] (X = halide ion) (**Figure 7**). Ni(II) pyridine complexes offer easy preparation in the organic solvent by the combination of nickel salt and pyridine and are stable against aerial oxidation, but Ni(I) pyridine complexes are sensitive to air and moisture. Though nickel can be coordinated up to six pyridines the stability of these complexes are very low [56].

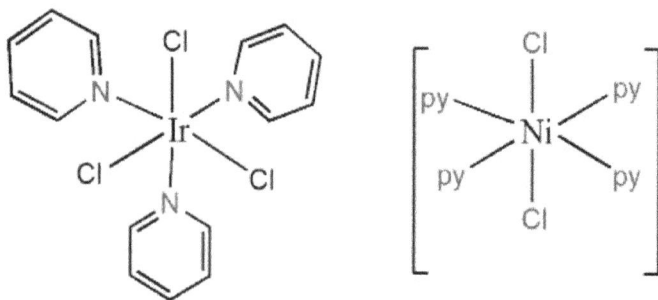

Figure 7. Representative pyridine complexes of Ir and Ni.

The palladium and platinum have many similarities in complex formation. The main oxidation states of these two metals are +II and +IV, yet pyridine complexes are known with Pd(0), Pt(0) Pd(I) and Pt(I) oxidation states. All other pyridine complexes are mostly square planar, though six coordinated complexes are common in +4 oxidation states. Majority of the complex contain one or two pyridine ligand in metal coordination sphere. Tetra pyridine complexes such as [Pd(py)$_4$](PF$_6$)$_2$, [57], [Pt(py)$_4$Br$_2$] [58] were also synthesized. The examples of four coordinated complexes are [Pd(py)$_2$Cl$_2$] [59] (Figure 8), [Pd(py)Br$_2${NH(CH$_2$CH$_2$NH$_2$)$_2$}] [60], [Pt(py)(Me)$_2$(CH$_3$COO)] [61], [Pt(py)$_2$X$_2$] [62] and so on, the Pd(IV) and Pt(IV) complexes [63–65] are [Pd(py)$_4$Cl$_2$]$^{2+}$, [PdCl$_4$(py)$_2$], and [Pt(py)$_2$Me$_2$Cl$_2$] (Figure 8).

The palladium-pyridine complexes were derived from PdCl$_2$, K$_2$[PdCl$_4$], which reacts with pyridine in the organic solvent or with neat pyridine to provide Pd(II) complexes. A similar procedure was adopted to yield platinum complexes. PtCl$_2$, K$_2$[PtCl$_4$], and Zeiss salt were employed in alcoholic, DMSO, and the aqueous medium to combine with pyridine. [Pt(py)$_2$Cl$_2$] could be synthesized from K$_2$[PtCl$_4$]. Pt(IV) complexes are not derived directly from Pt(IV) starting complex such as K$_2$[PtCl$_6$], whereas Pt(II) pyridine complexes are oxidized using appropriate oxidizing agents. Hydrogen peroxide could oxidize [Pt(py)$_2$Cl$_2$] to [Pt(py)$_2$(OH)$_2$Cl$_2$]. The synthesis of Pd and Pt pyridine complexes are governed by trans-effect and the resultant complexes are aptly substituted product on a starting material by pyridine according to this effect.

Both copper and silver are capable to form complexes in three oxidation states (I–III), whereas gold only displays two oxidation states (I and III). The isolated pyridine complexes are either two, three, or four coordinated for +1 state such as [Cu(py)(PPh$_3$)$_2$] [66]. But Cu(II), Au(III) complexes could extend their coordination number to six. Among this Cu(II) pyridine complexes are extensively studied. It can have the varied number of pyridine ligand in coordination sphere. Examples are [Cu(py)$_2$(ox)] [67], [Cu(py)$_3$(NCS)$_2$] [68], [Cu(py)$_4$(H$_2$O)$_2$]$^{2+}$ [69] (Figure 8) and so on. Complexes with six pyridine ligands, [Cu(py)$_6$]Br$_2$ [70] is also reported. It was prepared by blending CuBr$_2$ with pyridine in ethanol medium. The compound is stable but readily decompose in presence of moisture resulting [Cu(py)$_2$Br$_2$].2H$_2$O

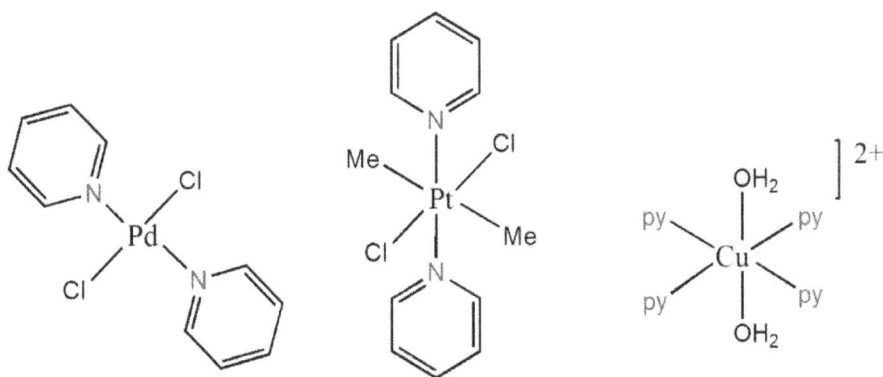

Figure 8. Representative pyridine complexes of Pd, Pt and Cu.

Ag(I) pyridine complexes are usually unstable and difficult to isolate. The coordination number, usually two or three, is another constraint to grow interesting chemistry with silver-pyridine complexes. Ag(I)-pyridine complexes preparation involve reaction of pyridine with Ag(I) salt in the solvent or in neat reactants. Often water remains a preferred solvent for such preparation. The examples include [Ag(py)(SCN)] [71], [Ag(py)(CN)] [72] (**Figure 9**), [Ag(py)$_4$]NO$_3$ [73], and so on. When the picolinic acid (picH) reacted with Ag(I) salt it resulted in a four-coordinated Ag(I) complex of formula [Ag(picH)(pic)] [74]. Ag(II)-pyridine complexes are usually derived from oxidation of Ag(I) counterpart using suitable oxidizing agents. Crystal structure of Ag(II)-bis(pyridine-2,3-dicarboxylate) shows a square planar geometry [75], where silver(II) is coordinated through pyridine nitrogens and two oxygen atoms of carboxylate groups at 2-position.

Compare to silver, gold has more numerous pyridine complexes. It's pyridine chemistry evolves with Au(I) and Au(III) states where pyridine and its various derivative were employed to derive desired complexes. For instance, Au(I) pyridine complexes could be achieved from the reaction of bis(acetonitrile) gold(I) perchlorate and 2-, 3- or 4-cyanopyridine in carbon tetrachloride. The corresponding cyanopyridine was used in the excess amount to produce [Au(n-CN-py)]ClO$_4$ (n = 2, 3, 4) [76]. Au (III) complexes could be prepared from Au(III) starting materials. The common starting materials used are AuCl$_3$ or HAuCl$_4$. [Me$_2$Au(py)X] (X = halide/pseudohalide) (**Figure 9**) type complexes were obtained by blending [Me$_2$AuCl]$_2$ and stoichiometric pyridine in cyclopentane at room temperature [77].

2.3. Application of metal-pyridine complexes

Transition metal pyridine complexes have proved their importance in various applications in different fields. Metal complexes with mixed ligand environment were studied and their applications are reported. Only a few representative complexes and their applications are mentioned here (**Figure 10**). Titanium pyridine complexes of the type [Ti(py)$_x$Cl$_y$] were explored as alkene and alkyne polymerization catalysts. These titanium complexes were used along with various aluminum cocatalyst such as RAlCl$_2$, R$_2$AlCl, and R$_3$Al (R = alkyl groups) as described in Ziegler-Natta catalytic process [78–80]. Pyridine complexes derived from VCl$_3$, VCl$_4$, and VOCl$_3$ act as olefin polymerization catalyst.

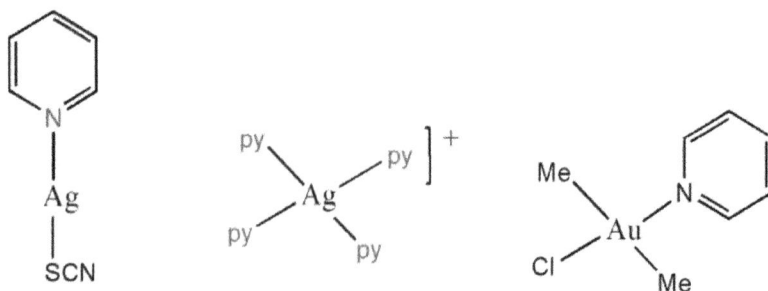

Figure 9. Representative pyridine complexes of Ag, and Au.

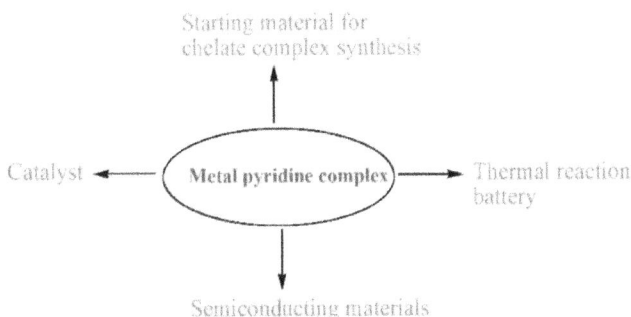

Figure 10. Application of metal pyridine complexes.

Mn(III)-pyridine complexes were found useful in developing photographic images [81]. The Cu (NO$_3$) complex with pyridine was found as a useful semiconductor for thermistors [82] and was tested to explore as an explosive [83]. [Mn(py)$_4$Cl$_2$] was applied in the thermal reaction battery, where a complex is thermally decomposed into a conducting salt, which forms an electrolyte [84]. The [ReCl$_4$(NO)(py)$_2$] complex showed catalytic activity in the hydrogenation of cyclohexene [85]. Fe(II)-pyridine complexes acts catalyst along with sodium borohydride in the process of selective reduction of nitrobenzene to phenylhydroxylamine [86]. Pyridine ligands are often labile and can be replaced with chelating ligands such as ethylenediamine, oxalate and so on. Fe(II) pyridine complexes used as starting material for synthesis of Fe(II) chelate complexes without pyridine ligands [87]. [Ru(bpy)$_2$(NO)(py)]$^+$ catalyzes the electrochemical oxidation of triphenylphosphine [88], [Ru(py)$_{6-n}$X$_n$] type complexes have a significant amount of π-back donation to the pyridine. This makes the coordinated pyridine suitable for electrophilic substitution. In [RuCl$_3$(py)$_3$], pyridine undergoes nitration to give 3-nitropyridine at elevated temperature [89]. Several Rh(I) and Rh(III) pyridine complexes catalyse hydroformylation of ethylene hydrocarbons [90, 91]. A complex, [RhCl$_2$(BH$_4$)(DMF)(py)$_2$] can act as the homogeneous catalyst for hydrogenation of pyridine to piperidine [92].

Early transition metal pyridine complexes are also explored in catalytic synthesis of amine and N-heterocycles. Quite a few mono (2-aminopyridinate)tris(dimethylamido) titanium complexes were synthesized and explored for intramolecular hydroamination reactivity using aminoalkene substrates. A titanium catalyst capable of room-temperature hydroamination reactivity was identified for the synthesis of *gem*-disubstituted 5- and 6-membered-ring products [93].

A series of 3-substituted-2-pyridonate ligands were employed with titanium and used as ancillary ligands for targeting selectively intramolecular hydroaminoalkylation over hydroamination. It was found that bis(3-phenyl-2-pyridonate)bis(dimethylamido) titanium complex is selective for hydroaminoalkylation over hydroamination. This can selectively α-alkylate primary aminoalkenes with marked substrate-dependent diastereoselectivity [94].

2.4. Theoretical look at metal pyridine complexation

The above discussion brings out the fact that pyridine ligation to metal studied in a mixed ligand environment. Synthesis of "pure" pyridine complexes of the type $[M(py)_4]^{n+}$ and $[M(py)_6]^{n+}$ are difficult. Experimentally sequential addition of pyridine to a particular metal center to its highest coordination number is impractical. Thus theoretical approach adopted to study pyridine binding to metals using pyridine ligands to a group of first transition metals ions in their +2 oxidation states. DFT calculation employed by Rodgers [95] to determine the ground state structure and sequential binding energies of $[M^{2+}(Py)_x]$ complexes, where x = 1–6 and M = Fe^{2+}, Co^{2+}, Ni^{2+}, Cu^{2+}, and Zn^{2+}. On gradual addition of pyridine, the metal ion adopts different geometries and the M-Py bond lengths and py-M-py angles got optimized according to the geometrical arrangements. $[M^{2+}(py)_4]$ and $[M^{2+}(py)_5]$ usually adopt distorted tetrahedral and distorted trigonal bipyramidal geometry. $[Cu^{2+}(py)_4]$ exhibits significant distortion from an ideal tetrahedral geometry to close to square planar arrangement, whereas $[Ni^{2+}(py)_5]$ and $[Cu^{2+}(py)_5]$ complexes adopt a square pyramidal geometry. The change of bond distances and bond angles evaluated on the addition of pyridine ligand to the metal are summarized in **Table 1** for $[M^{2+}(py)_x]$ (M^{2+} = Fe^{2+} & Cu^{2+}). In a trend M^{2+} – N bond distances decrease from Fe^{2+}(1.939 Å) to Co^{2+}(1.854 Å) to Ni^{2+}(1.848 Å) to Cu^{2+}(1.844 Å), and then slightly increase for Zn^{2+}(1.870 Å), which is expected on the basis of the sizes of these cations and, consequently, stronger electrostatic contributions to the binding. For a particular M^{2+} center, the bond lengths gradually increase with the increased number of pyridines in the coordination sphere. The Fe-Py distance 1.939 Å in $[Fe^{2+}(py)]$ increases to 2.302(6) in $[Fe^{2+}(py)_6]$. This bond distance enhancement is due to increase in steric crowding around the metal center.

Complex	M^{2+} – N (Å)	∠$NM^{2+}N$ (deg)
Fe^{2+}(Py)	1.939	—
Fe^{2+}(Py)$_2$	1.953 (2)	179.9
Fe^{2+}(Py)$_3$	2.019 (3)	120.0(3)
Fe^{2+}(Py)$_4$	2.079 (4)	108.7 (2), 111.1 (4)
Fe^{2+}(Py)$_5$	2.161 (3), 2.229 (2)	86.5 (2), 90.2 (2), 94.7 (2), 111.1 (2), 137.7, 170.7
Fe^{2+}(Py)$_6$	2.302 (6)	89.3 (6)
Cu^{2+}(Py)	1.844	—
Cu^{2+}(Py)$_2$	1.877 (2)	180.0
Cu^{2+}(Py)$_3$	1.910 (2), 1.929	104.8(2), 150.5
Cu^{2+}(Py)$_4$	2.004 (4)	92.8, 97.2(3), 141.9(2)
Cu^{2+}(Py)$_5$	2.065, 2.277 (4)	86.6 (2), 90.2 (4), 96.4 (2), 104.5(2), 159.1(2)
Cu^{2+}(Py)$_6$	2.027 (4), 3.174(2)	89.5 (6), 90.5(6), 179.0 (3)

Values are reproduced from ref. [95].

Table 1. Theoretical bond angle and bond distances of $[M^{2+}(Py)_x]$ (M^{2+} = Fe^{2+} & Cu^{2+}).

Complex	Binding energy (kJ/mol)	Complex	Binding energy (kJ/mol)
$Fe^{2+}(py)$	621.8	$Cu^{2+}(py)$	916.8
$Fe^{2+}(py)_2$	424.9	$Cu^{2+}(py)_2$	395.3
$Fe^{2+}(py)_3$	302.4	$Cu^{2+}(py)_3$	252.8
$Fe^{2+}(py)_4$	170.3	$Cu^{2+}(py)_4$	147.5
$Fe^{2+}(py)_5$	68.1	$Cu^{2+}(py)_5$	65.4
$Fe^{2+}(py)_6$	57.3	$Cu^{2+}(py)_6$	41.8

Values are taken from ref. [95].

Table 2. Sequential binding energy of $[M^{2+}(py)_x]$ (M^{2+} = Fe^{2+} and Cu^{2+}) complexes at 0 K.

The bonding of M^{2+} cations to the nitrogen lone pairs facilitates strong ion-dipole and ion-induced dipole interactions. This is evident from M^{2+} – N bonds orientation along the dipoles of the pyridine ligands. The M^{2+} cation binds the nitrogen lone pair forming M-py sigma bond. The dominant charge transfer involves ligand-to-metal sigma donation from nitrogen lone pairs of pyridine to the vacant valence shell of the metal and the metal-to-ligand charge transfer via π back bonding to the unoccupied π^* orbitals of pyridine ligand.

The strength of binding of the pyridine ligands to the M^{2+} cations decreases gradually with increasing ligation around the metal cation. This decrease in sequential bonding energy contributed to decreased attractive electrostatic interactions, charge transfer from the pyridine ligands to the metal cation, Jahn–Teller distortion, and ligand–ligand repulsion. Binding of the first ligand is comparatively strong as electrostatic interactions and charge transfer are both important contributors to the bonding in the $M^{2+}(py)x$ complexes. The binding energy of the second, third, fourth, fifth, and sixth ligands decrease on sequential ligation of pyridine because the effective charge retained by the metal center decreases, and the M^{2+} – N bond distances increase. The extent of bonding is expected to increase from Fe^{2+} to Co^{2+} to Ni^{2+} to Cu^{2+} following the decrease of ionic radii of the metal cations. The trend is observed as Cu^{2+}-py binding energy (916.8 kJ) in $[Cu^{2+}(py)]$ is greater than the same (621.8 kJ) of $[Fe^{2+}(py)]$ as illustrated in **Table 2**.

3. Conclusion

In this chapter, the capability of pyridine as a ligand to transition metal ions was discussed. It has wealth of orbitals which are utilized in the formation of the bond to metal centers. Both electrostatic interactions and charge transfer are important factors in the bonding of pyridine to metals. The prominent charge transfer interactions make ligand-to-metal σ donation. The metal-pyridine bonding further got boosted by metal-to-ligand charge transfer through π back donation from metal $d\pi$ orbitals to the unoccupied π^* orbitals of the pyridine ligand(s). The pyridine coordinates to all the transition metals known in the periodic table and hence has a rich literature of chemistry. Most of the complexes accommodate pyridine in a mixed ligand atmosphere though $[M(py)_x]^{n+}$ type complexes were also characterized. Theoretical studies showed that the stability of pyridine complexes decreases with increasing ligation of pyridines around the metal center. A range of pyridine complexes was found their applications in numerous fields.

Acknowledgements

The author gratefully acknowledges the contribution of Prof. Samudranil Pal, University of Hyderabad and Prof. Anadi Charan Dash, Retired professor of Utkal University in preparing this chapter.

List of abbreviation and symbols

pm	picometer
Å	Angstrom
py	pyridine
bpy	bipyridine
SCN	thiocyanate
NCS	isothiocyanate
2-Me-py	2-methyl pyridine
4-Me-py	4-methyl pyridine
4-Et-py	4-ethyl pyridine
2, 6 –Me$_2$-py	2, 6 dimethyl pyridine
acac	acetylacetone
ox	oxalate

Author details

Satyanarayan Pal

Address all correspondence to: snpal75@gmail.com

Department of Chemistry, Utkal University, Bhubaneswar, India

References

[1] Joule JA, Mills K. Heterocyclic Chemistry. 5th ed. Chichester: Blackwell Publishing; 2010

[2]   Huheey JE, Keiter EA, Keiter RL. Inorganic Chemistry. 4th ed. New York: Herper Collins Publisher; 1993

[3]   Firsova NL, Kolodyazhnyi YV, Osipov OA. Zhurnal Obshchei Khimii. 2151;**1969**:39

[4]   Shibata S. Analytica Chimica Acta. 1963;**28**:388

[5]   Matignon C. Annales de Chimie Physique. 1906;**8**:433

[6]   Dutt NK, Sen Gupta AK. Zeitschrift für Naturforschung. 1975;**30b**:769

[7]   Hnilickova M, Sommer L. Zeitschrift für Analytische Chemie. 1960;**177**:425

[8]   Sas TM, Komissarova LN, Arnatskaya NI. Zhurnal Neorganicheskoi Khimii. 1971;**16**:87

[9]   Zikmund M, Valent A, Stepnickova L. Chemicke Zvesti. 1967;**21**:901

[10] Zikmund M, Foniok R, Valent A. Chemicke Zvesti. 1965;**19**:854

[11] Bmeleus HJ, Rao GS. Journal of the Chemical Society. 1958:4245

[12] Fowles GWA, Russ BJ, Willey GR. Journal of the Chemical Society, Chemical Communications. 1967;**646**

[13] Hieber W, Peterhans J, Winter E. Chemische Berichte. 1961;**94**:2572

[14] Torii Y, Iwaki H, Inamura Y. Bulletin of the Chemical Society of Japan. 1967;**40**:1550

[15] Seifert HJ, Auel T. Journal of Inorganic and Nuclear Chemistry. 2081;**1968**:30

[16] Khamar MM, Larkworthy LF, Patel KC, Phillips DJ, Beech G. Australian Journal of Chemistry. 1974;**27**:41

[17] Golub AM, Kostrova RA. Dopov. Akad. Nauk Ukr. SSSR. 1963;**1061**. [Chemical Abstracts. 1964;**60**:3704]

[18] Miles SJ, Wilkins JD. Journal of Inorganic and Nuclear Chemistry. 2271;**1975**:37

[19] Machin DJ, Sullivan JF. Journal of the Less-Common Metals. 1969;**19**:405

[20] Pfalzgraf LG, Riess JG. Bulletin De La Societe Chimique De France. 1968:4348

[21] Osipov OA, Kashireninov OE, Leschenko AV. Zhurnal Neorganicheskoi Khimii. 1964;**9**:734

[22] Allbutt M, Feenan K, Fowles GWA. Journal of the Less-Common Metals. 1964;**6**:299

[23] Hubert-Pfalzgraf LG, Guion J, Riess JG. Bulletin de la Société Chimique de France. 1971:3855

[24] Herberhold M, Alt H. Justus Liebigs Annalen der Chemie. 1976;**292**

[25] Relan PS, Bhattacharya PK. Journal of the Indian Chemical Society. 1969;**46**:534

[26] Majumdar MN, Saha AK. Journal of Inorganic and Nuclear Chemistry. 1976;**38**:1374

[27] McCain DC. The Journal of Physical Chemistry. 1975;**79**:1102

[28] Oh SO. Daehan Hwahak Hwoejee. 1968;**12**:93. [Chemical Abstracts. 1969;**70**:43547w]

[29] Park DW, Sub OT. Taehan Hwahak Hoechi. 1975;**414**:19. [Chemical Abstracts. 1976;**84**: 188795z]

[30] Seifert HJ, Petersen F, Woehrmann H. Journal of Inorganic and Nuclear Chemistry. 2735;**1973**:35

[31] Tebbe FN, Muetterties EL. Inorganic Chemistry. 1968;**7**:172

[32] Fyfe WS. Journal of the Chemical Society. 1950:790

[33] Das AK, Rao DVR. Zeitschrift für Anorganische und Allgemeine Chemie. 1970;**379**:213

[34] Potter WC, Taylor LT. Inorganic Chemistry. 1976;**15**:1329

[35] Atwood JD, Brown TL. Journal of the American Chemical Society. 1976;**98**:3155

[36] Uguagliati P, Zingales F, Trovati A, Cariati F. Inorganic Chemistry. 1971;**10**:507

[37] Lock CJL, Guest A. Canadian Journal of Chemistry. 1971;**49**:603

[38] Sur B, Sen D. Science and Culture (Calcutta). 1960;**26**:85. [Science and Culture. 1961;**55**:1263]

[39] Hieber W, Lux F, Herget C. Zeitschrift für Naturforschung. 1965;**20b**:1159

[40] Kuzina AF, Oblova AA, Spitsyn VI. Zhurnal Neorganicheskoi Khimii. 2630;**1972**:17

[41] Doedens RJ, Dahl LF. Journal of the American Chemical Society. 1966;**88**:4847

[42] Hieber W, Sonnekalb F. Berichte. 1928;**61B**:2421

[43] Laure TA. Analytica Chimica Acta. 1968;**40**:437

[44] Keene FR, Salmon DJ, Meyer TJ. Journal of the American Chemical Society. 2384;**1977**:99

[45] Charonnat R. Annali di Chimica. 1931;**16**:123

[46] Soucek J, Vrestal J. Collection of Czechoslovak Chemical Communications. 1961;**26**:1931

[47] Koda Y. Nagoya Kogyo Gijutsu Shikensho Hokoku. 1966;**15**:155 [Chemical Abstracts. 1968;**69**:48876s]

[48] Yih CC, Deeming AI. Journal of the Chemical Society Dalton Transactions. 1975;**2091**

[49] Buckingham DA, Dwyer FP, Goodwin HA, Sargeson AM. Australian Journal of Chemistry. 1964;**17**:315

[50] Liptay G, Burger K, Mocsari-Fulop E, Porubszky I. Journal of Thermal Analysis. 1970;**2**:25

[51] Goodgame M, Hayward PJ. Journal of the Chemical Society A. 1971;**3406**

[52] Addison AW, Dawson K, Gillard RD, Heaton BT, Shaw H. Journal of the Chemical Society Dalton Transactions. 1972:589

[53] Fujiwara S, Watanabe T, Inoue T. Chemistry Letters. 1974;**755**

[54] Bennett MA, Mitchell TRB. Inorganic Chemistry. 2396;**1976**:15

[55] Kutner W, Galus Z. Electrochimica Acta. 1975;**20**:301

[56] Ablov AV, Nazarova LV. Zhurnal Neorganicheskoi Khimii. 1957;**2**:53

[57] Trofhnenko S. Inorganic Chemistry. 1973;**12**:1215

[58] Goremykin VI, Gladyshevskaya KA. Izv. Akad. Nauk SSSR, Ser. Khim. 1943;**338**

[59] Panasyuk VD, Denisova TI. Izvestiya Akademii Nauk SSR, Seriya Fizicheskaya. 2517;
1971:16

[60] Basolo F, Gray HB, Pearson RG. Journal of the American Chemical Society. 1960;**82**:4200

[61] Kuyper J, Vrieze K. Transition Metal Chemistry (Weinheim). 1976;**1**:208

[62] Irving RJ, Magnusson EA. Journal of the Chemical Society. 1957;**2018**

[63] Eaborn C, Farrell N, Murphy JL, Pidcock A. Journal of the Chemical Society Dalton
Transactions. 1976;**58**

[64] Rosenheim A, Maass TA. Zeitschrift für Anorganische und Allgemeine Chemie. 1898;
18:331

[65] Hall JR, Swile GA. Australian Journal of Chemistry. 1971;**24**:423

[66] Reichle WT. Inorganica Chimica Acta. 1971;**5**:325

[67] Lapiere C. Journal de Pharmacie de Belgique. 1948;**17**:3 [Chem. Abstr. 1948;42:3754]

[68] Calzolari J. Berichte. 1911;**43**:2217

[69] Leussing DL, Gallagher PK. The Journal of Physical Chemistry. 1960;**64**:1631

[70] Allan JR, Brown DH, Nuttall RH, Sharp DWA. Journal of the Chemical Society A.
1966:1031

[71] Mathews JH, Kraus EL, Bohnson LV. Journal of the American Chemical Society.
1917;**39**:358

[72] Varet R. Comptes Rendus de l'Académie des Sciences. 1891;**112**:390

[73] Wilke-Doerfurt E, Balz G. Zeitschrift für Anorganische und Allgemeine Chemie. 1927;
159:197

[74] Deloume JP, Faure R, Loiseleur H. Acta Crystallographica. 2709;**1977**:B33

[75] Drew MGB, Matthews RW, Walton RA. Inorganic and Nuclear Chemistry Letters.
1970;**6**:277

[76] Junior FF, Iwamoto RT. Inorganic Chemistry. 1965;**4**:844

[77] Stocco GC, Tobias RS. Journal of the American Chemical Society. 1971;**93**:5057

[78] British Patent. 1962;**891**:646. [Chemical Abstracts. 1962;**57**:2424]

[79] British Patent. 1963;**940**:137. [Chemical Abstracts. 1965;**62**:10546]

[80] Berlin AA, Cherkashin MI, Kisilitsa PP. Izv. Akad. Nauk. SSSR, Ser. Khim. 1965;**1875**:81

[81] Jonge J D, Jonker H, Keuning K. Dippel C. J: US Patent. 1956; **2**: 764, 484. [Chemical Abstracts. 1957;**51**:3334]

[82] French Pat. Addn. 1967;**89**:238. [Chemical Abstracts. 1968:**69**:47521k]

[83] Cirulis A, Straumanis M. Journal für Praktische Chemie. 1943;**162**:307

[84] Bald JF Jr. US Patent. 1974;**3**:811-948. [Chemical Abstracts. 1974;**81**:66200u]

[85] Ryashentseva MA, Minachev KM, Belanova EP, Shapiro ES, Ovchinnikova NA. Izv. Akad. Nauk SSSR, Ser. Khim. 1976;**2647**

[86] Arai Y, Mijin A. Kitami Kogyo Tanki Daigaku, Kenkyu Hokoku. 1973;**5**:69. [Chemical Abstracts 1974;**80**:47571d]

[87] Constant LA, Davis DG. Journal of Electroanalytical Chemistry and Interfacial Electrochemistry. 1976;**74**:85

[88] Keene FR, Salmon DJ, Meyer TJ. Journal of the American Chemical Society. 1977;**99**:4821

[89] Soucek J. Collection of Czechoslovak Chemical Communications. 1962;**21**:1645

[90] French Patent. 1968;**1**:549, 414. [Chemical Abstracts. 1970;**72**:2995p]

[91] Stone FGA, Mukhedkar AJ, Mukhedkar VA, Green M. Journal of the Chemical Society A. 1970;**3158**

[92] McQuilh FJ, Abley P, Jardine I. Journal of the Chemical Society C: Organic. 1971;**840**

[93] Chong E, Qayyum S, Schafer LL, Kempe R. Organometallics. 2013;**32**:1858

[94] Chong E, Schafer LL. Organic Letters. 2013;**15**:6002

[95] Nose H, Rodgers MT. The Journal of Physical Chemistry. A. 2014;**118**:8129